Dr. Emily Dawson provides a much-needed perspective on science education, pointing out the issues that are difficult to grasp or even see from our privileged vantage points.

Marianne Achiam, *Department of Science Education, University of Copenhagen*

Issues of equity and exclusion are vital considerations for informal science learning research, policy and practice. This book offers a powerful and much-needed analysis based, crucially, on the views and experiences of those who are usually excluded from settings such as science museums and centres due to inequalities of 'race', social class and gender. A 'must read' for anyone concerned with equity and inclusion in informal science learning.

Louise Archer, *Karl Mannheim Professor of Sociology of Education, UCL Institute of Education*

Equity, Exclusion and Everyday Science Learning

Equity, Exclusion and Everyday Science Learning explores how some people are excluded from science education and communication. Taking the role of science in society as a starting point, it critically examines the concept of equity in science learning and develops a framework to support inclusive change.

This book presents a theoretically informed, empirically detailed analysis of how people from minoritised groups in the UK experience science and everyday science learning resources in their daily lives. The book draws on two years of ethnographic research carried out in London with five community groups who identified as Asian, Somali, Afro-Caribbean, Latin American and Sierra Leonean. Exploring their experiences of everyday science learning from a sociological perspective, with social justice as a guiding concern, this book opens with a theory of exclusion and closes with a theory of inclusion.

Equity, Exclusion and Everyday Science Learning is not only an essential text for postgraduate students and postdoctoral researchers of Science Education, Science Communication and Museum Studies, but for any professional working in museums, science centres and institutional public engagement.

Emily Dawson is an Associate Professor in the Department of Science and Technology Studies at University College London. Her work focuses on how people engage with and learn about science, with an emphasis on equity, in particular the construction of publics and 'non' publics for science, and the role of privilege in such processes.

Routledge Research in Education

This series aims to present the latest research from right across the field of education. It is not confined to any particular area or school of thought and seeks to provide coverage of a broad range of topics, theories and issues from around the world.

Recent titles in the series include:

For a complete list of titles in this series, please visit: www.routledge.com/Routledge-Research-in-Education/book-series/SE0393

Equity, Exclusion and Everyday Science Learning

The Experiences of Minoritised Groups

Emily Dawson

LONDON AND NEW YORK

First published 2019
by Routledge
2 Park Square, Milton Park, Abingdon, Oxon OX14 4RN

and by Routledge
52 Vanderbilt Avenue, New York, NY 10017

First issued in paperback 2020

Routledge is an imprint of the Taylor & Francis Group, an informa business

British Library Cataloguing-in-Publication Data
A catalogue record for this book is available from the British Library

Library of Congress Cataloging-in-Publication Data
Names: Dawson, Emily, author.
Title: Equity, exclusion and everyday science learning : the
    experiences of minoritised groups / Emily Dawson.
Description: Abingdon, Oxon ; New York, NY : Routledge, 2019. |
    Includes bibliographical references.
Identifiers: LCCN 2018041514 | ISBN 9781138289949 (hardback) |
    ISBN 9781315266763 (ebook)
Subjects: LCSH: Science—Social aspects.
Classification: LCC Q125 .D2935 2019 | DDC 507.1—dc23
LC record available at https://lccn.loc.gov/2018041514

ISBN 13: 978-0-367-66215-8 (pbk)
ISBN 13: 978-1-138-28994-9 (hbk)

Typeset in Galliard
by Apex CoVantage, LLC

This book is for my parents, Françoise Dawson
and Iain Dawson.

Contents

Figures

Preface and acknowledgements

This book is still not quite finished in my head, but it eventually had to become finished on the page, and I want to briefly explain why I wrote it. I came to this research from a science museum, science documentaries and hands-on science communication background. I organised this research in order to try to make sense of what I was encountering at work: the prejudices, the privilege, the unspoken assumptions about audiences and staff recruitment, the foregone conclusions about representation.

This book stands, I hope, as a testament to the stories and experiences of the participants who took part in this project. It remains my great honour to have worked with 59 people from five grass-roots community groups (an Afro-Caribbean group, a Somali group, a Sierra Leonean group, a Latin American group and an Asian group). For obvious reasons I cannot list their names here, but their perspectives, insights and experiences form the core around which this book is organised. They shared their time, opinions, experiences, food and laughter so generously. What I learnt from them shocked me. I knew issues of inclusion and representation were bad when it came to staff and visitors, users or audiences in the spaces where I worked. But exclusion, inequality and inequity were more embedded in participants' everyday science learning experiences than I could have imagined. I hope this book does justice to the experiences they shared with me.

In pulling together all of the research described in this book I am reminded of a colleague who commented one day that he felt I had 'stitched up' science museums. I have spent the majority of my adult life working in one way or another across the various fields involved in everyday science learning. Although I have no desire to stitch anyone up, I also have no desire to sugar-coat how exclusion operates within these fields. If we are to stand a chance of reimagining everyday science learning such that meaningfully inclusive, equitable practice becomes the norm, then we must face the challenges of exclusion and equity head on. We accomplish nothing by pretending there are no problems and this deception would belittle participants' experiences in ways I am not prepared to accept. As such, I have written about exclusion and, as I discuss at the end of the book, inclusion, with the hope that the data and arguments presented here support reflection and discussion across everyday science learning practices and, ideally, lead to more inclusive practice.

Before the book goes any further I must thank more people; please indulge me or skip ahead. In addition to the participants thanked previously who were directly involved in the research this book is based on, I want to thank and recognise the contributions of colleagues near and far whose work has supported, challenged and developed my own. I confess I did write out the long list of all their names, but it became faintly ridiculous, so I will just say thank you again and hope that you know how much your support has meant to me.

Specifically I want to thank colleagues at the Science Communication Unit at the University of the West of England for their support when I applied for funding for a PhD, particularly Emma Weitkamp, Karen Bultitude and Clare Wilkinson. I want to thank fellow students, supervisors and later colleagues at the department where I carried out my PhD research, then called the Department of Education and Professional Studies at King's College London, particularly Jonathan Osborne, Justin Dillon and Anwar Tlili for their supervision during my PhD. Thanks also to the UK's Economic Social and Research Council who funded the PhD project this book grew from (award reference: ES/G018448/1).

I want to thank colleagues at the Department of Science and Technology Studies at UCL, where I wrote this book and whose friendship, enthusiasm and tangible support (for instance, in the form of sharing book proposals!) I have greatly appreciated. Thanks also to Lori Coletti Campbell and Joe Cain for their support during a hideous miscarriage and an unpleasant third pregnancy, and for encouraging me to take a research term to write this book once I returned to work. I want to thank Billy Wong for his encouragement at the start of this project – he made a book seem possible in a way I had not really considered. I also want to thank colleagues with whom I have worked on two research projects whose themes echo those of this book, the Youth Equity and Science, Technology, Engineering and Mathematics (YESTEM) project and the Enterprising Science project. I want to thank them not only for their continued support as this book took shape and I became increasingly distracted, but also for being constant sources of inspiration, from whom I learn so much.

Each chapter in this book was read and fed back on by at least one academic and one practitioner colleague from across the varied worlds of science communication, public engagement, science education, informal science learning and museum studies. In alphabetical order then I must thank Amy Seakins, Effrosyni Nomikou, Heather King, Heather Mendick, Imran Khan, Ivvet Modinou, Jo Bryant, John Falk, Katherine Cecil, Louise Archer, Mat Hickman, Melissa Glackin, Sai Pathmanathan, Spela Godec, Steve Cross, Subhadra Das, Toni Dancstep and Vanessa Mignon for their honest feedback. I must also thank Simon Lock for spending a lot of time discussing the relative merits and disadvantages of the term 'everyday science learning'. As he says, it will do for now. My gratitude for their kindness in helping me with this work is enormous – thank you all. But for the under-developed ideas, mistakes and confusing sentences I take full responsibility.

Part of the reason I wrote a book is that I greatly value other books when it comes to challenging my ideas, learning from others and exploring how other people think. In the REF regime that structures contemporary academic life in the UK, where papers are the dominant unit of analysis, books seem increasingly like spaces to be cherished. I am grateful to those authors whose writing has given me the tools (both theoretically and in terms of language) to think and write about my own work. Particular thanks must go to Nirmal Puwar, Richard Sandell and Valerie Walkerdine. I do not cite all of their work in this book, but I know their writing shaped my thinking.

I must also thank Shelly Pike, Debbie Neenan, Yvonne Knight and Chelcie Ashford, who looked after my daughter while I wrote this book. It meant the world to me that I knew she was having a great time with them while I typed, rewrote, read, edited, thought, changed my mind and typed some more. I also want to thank Anna Bobak whose care for our home and family has been such a blessing. I am deeply grateful to these women, whose hard work makes my work possible.

Last, but certainly not least, I am grateful to my friends and family, without whose patience, support and enthusiasm over the past 10 years I am quite sure I would never have started a PhD, let alone finished this book. In addition to those friends who literally supported the writing of this book by reading drafts I want to thank Alice Marsh, Anna Chapman Andrews, Jonathan Shakovskoy and their family, Berry Rose, Joe Kelly and their family, Cathy O'Brien, Charlie Browne, Ellie and Gareth Estchild, Helen Featherstone, Jess Floyd, Jo Hargreaves, Karl Marrow and their family, Laura Perry, Charlie Turner and their family, Morgan Lloyd Malcolm, Natalie Witting, Tamsyn Dent, Paddy Chatterton and their family and Victoria Ansell. For the same reasons I must thank Pierre Dawson, Gemma Elford-Dawson, Ben Cook, Angie Cook, Vivian Cook and Chris Cook.

The thanks I owe to my parents, Françoise Dawson and Iain Dawson, are such that I am not sure I can articulate them, except to repeat thanks and more thanks. During the final stages of writing this book my mother sent me a postcard with the words "long may you promote fairness and equality in these dry subjects" – she is a staunch feminist, has lived most of her life away from the country of her birth and is not a huge fan of science. My father, in comparison, worked his whole life in the sciences and gave me a life-long appreciation for how science works. I am keenly aware that without their support, the last few years would have been impossible. My mother moved in with us after I nearly died when my daughter was born. Without help from my parents our lives would have been much harder (and the bizarre paperwork associated with organising a book contract would likely have been on hold indefinitely). Now that I've finished it, I see that this book was in many ways their doing.

Finally, I want to thank my own little family of Tom Cook and Olwyn Dawson. Thank you for putting up with me in the academic-recluse mode that I switched into to get this book finished. Thanks for your generosity with time and attention when it came to discussing things like how best to represent theoretical concepts as diagrams. Thanks too for your excellent distractions. You are the best. I love you both.

Credits

Parts of this book have already been published in various guises already, all but one of them under open-access agreements. The parts of Chapter Four that tell Fatima's story are republished with permission of Taylor and Francis Group LLC Books from my chapter titled "When science is someone else's world", pp. 82–92 in *Intersections of Formal and Informal Science*, published in 2016, edited by Lucy Avraamidou and Wolf-Michael Roth; permission conveyed through Copyright Clearance Center, Inc.

All the other papers drawn on here are under open-access agreements. In order of chapter, here are their details:

Parts of Chapter Two have been published before as "Reframing social exclusion from: science communication: Moving away from 'barriers' towards a more complex perspective", published in 2014 in the *Journal of Science Communication* under a Creative Commons Attribution – Non Commercial – No Derivative Works 4.0 License, (https://creativecommons.org/licenses/by/4.0/).

Other parts of Chapters Two and parts of Seven are reprinted by permission of Taylor and Francis Ltd, from my paper "Equity in informal science education: developing an access and equity framework for science museums and science centres", pp. 209–247 in the journal *Studies in Science Education*, published in 2014 under a Creative Commons Attribution – Non Commercial – No Derivative Works 3.0 License. (https://creativecommons.org/licenses/by/3.0/).

Parts of Chapters Three and Five have been published before as "Reimagining publics and (non)participation: Exploring exclusion from science communication through the experiences of low-income, minority ethnic groups", pp. 1–15 in the journal *Public Understanding of Science,* under a Creative Commons Attribution – Non Commercial – No Derivative Works 4.0 License, (https://creativecommons.org/licenses/by/4.0/).

Parts of Chapter Six and other parts of Chapter Seven have been published before as "Not Designed for Us: How Science Museums and Science Centers Socially Exclude Low-Income, Minority Ethnic Groups", pp. 981–1008 in the *Journal of Science Education* in 2014, under a Creative Commons Attribution – Non Commercial – No Derivative Works 4.0 License, (https://creativecommons.org/licenses/by/4.0/).

Even more parts of Chapter Seven have been published before as "Social justice and out-of-school science learning: Exploring equity in science television, science clubs and maker spaces", pp. 539–547 in the *Journal of Science Education* in 2017, under a Creative Commons Attribution – Non Commercial – No Derivative Works 4.0 License, (https://creativecommons.org/licenses/by/4.0/).

Chapter 1

Introduction

Exploring exclusion

ABDOU: You know when you start thinking of going [out], the science museum or a museum, is the last place that you would even think of, that you would even consider on your list, even if you were doing a list of hundreds of places that you want to go.

EMILY: It wouldn't even be on the list?

ABDOU: Not on the list!

EMILY: So why is that?

ABDOU: You cannot connect with it. It's for those people that it matters to [. . .] the museum, or science itself.

What does it mean to feel that science museums and indeed that "science itself" are not for you? And why does that matter for our societies? As Abdou, a middle-aged Sierra Leonean man, argued in the interview extract above, science and science museums were "only for those people that it matters to". As the research carried out for this book shows, he was not alone in his views. Opportunities to interact with, learn about, speak back to and laugh about science, are marked by structural inequalities that mirror and reproduce social advantages and disadvantages. Understanding how this happens is crucial, as I argue throughout this book, if we are to build more inclusive, more equitable practices within and beyond science communication and education.

This chapter sets out the context of the research discussed in this book and is organised into four sections. The first discusses why it is important to think about the brokering practices between science and society from a sociological perspective, with social justice as a guiding concern. The second section outlines the key terms and concepts used throughout the book. The third presents the details of the research this book is based upon, and the final section maps out each chapter of the book so that readers can pick their way through it as they choose.

Science and society

Opportunities to learn about, engage with, question and critique science are increasingly important in contemporary societies. Science and technology are embedded in people's lives in ways that are socially, culturally and politically significant. This ranges from issues people grapple with daily, to societal decisions about the legislation of particular technologies, or revelations about scientific scandals and 'miracle breakthroughs' in the mass media (Bradu, Orquin, & Thøgersen, 2013; Jasanoff, 2007; Michael, 2006; Nelkin, 1995). Given the degree to which science and technology affect our lives, accessible and equitable opportunities to engage with science are important to equip people with the tools, skills and information to navigate contemporary life or enter science-related professions.

For many people school and the mass media remain the key contexts for encounters with science and scientific information (Ipsos MORI, 2014; Osborne & Dillon, 2007). While schools and the mass media will doubtless continue to represent important sites for science learning and engagement, a world of other opportunities exists alongside them, from activities in science festivals and visits to zoos, to political consultations on scientific issues and citizen science groups (Ballard, Dixon, & Harris, 2017; Bonney et al., 2009; Falk & Dierking, 2012; Lewenstein, 2015; Stilgoe, Lock, & Wilsdon, 2014). Activities within this broad church of science-related public practices have been found to provide participants with opportunities to engage with science in ways that are useful, inspiring and educational (Bell, Lewenstein, Shouse, & Feder, 2009; Phipps, 2010; Stocklmayer, Rennie, & Gilbert, 2010). Questions remain, however, about how accessible, inclusive and equitable such practices are.

Countries like the UK are home to people from around the world and, at the same time, are beset by right-wing politics that seek to render the 'race'/ethnicity of the dominant ethnic majority invisible, while discriminating against racialised groups (Bhopal, 2018). For instance, the breaking story of the summer of 2018 was the detention and deportation of British-Caribbean people, who were framed as illegal immigrants by the Conservative governments' Home Office because they were born in the Caribbean. This 'Windrush Scandal' saw families separated, homes and jobs lost, but little official response or change to Home Office policy (Gentleman, 2018, p. 1). Stories like this speak to the structural inequalities of contemporary life in the UK.

People in Britain continue to live amidst the legacies of colonialism, with all the material, cultural, political and educational biases that follow from such entrenched inequalities (Eddo-Lodge, 2018; Gilroy, 2002; Hall, 2012; Mirza, 1992). Globalisation has meant, as Doreen Massey (1994) described so beautifully, that the ebbs and flows of ideas, images, emails, people, politics, trade and money circulate in complicated ways that do not necessarily follow the same patterns or result in the same outcomes for the Global North and the Global South. The interplay of 'race'/ethnicity, class, gender and the other intersecting subjectivities of people's lives today, as before, warrant careful consideration and, more than that, demand respect, recognition and understanding (Benhabib, 2002; Cho, Crenshaw, & McCall, 2013; Fraser, 2003).

In London, where the research for this book took place, the ripples of colonialism and globalisation continue to play out across the many and varied communities of this city, from its old, venerable institutions to its fleeting, pop-up festivals. Whose cultures and languages are represented in museums and galleries? Who is registered to vote? Whose food is readily available in high-street shops in different neighbourhoods? And, why does any of this matter if you are trying to understand the systems at work behind who watches science documentaries, who buys tickets to go to a science-comedy show, who gets involved in grass-roots campaigns about local pollution and who visits science museums?

This book is a sociological account of what happens when people from racialised groups in the UK encounter science through the various public activities that mediate between people and science. In this book I refer to these activities as everyday science learning practices. To think about how people relate to everyday science learning practices seems impossible to me without thinking at the same time about broader sociological questions. Not least, questions about how certain educational, political and cultural practices reproduce social inequalities along fault lines of 'race'/ethnicity, class, gender and their intersections. As a result, in this book I build on international research about how people engage with science across the different facets of their lives, concentrating mainly on experiences outside or beyond the formal school system. I also draw on theories of social justice from political philosophy and sociological research to understand what it means for people from minoritised, and in particular in this case racialised groups, to not engage with everyday science learning resources.

Read one way this is a book about everyday science learning practices. I examine how these practices are marked by and reproduce structural inequalities based on 'race'/ethnicity, class, gender and other facets of people's intersecting subjectivities. More hopefully, this book also explores how everyday science learning practices might disrupt entrenched patterns of oppression. Read another way this book provides a case study of social reproduction. In this sense, the focus on everyday science learning is simply a context that frames how structural inequalities operate in education, culture and politics. This book therefore contributes not only to research and practice in science education, science communication, public engagement with science and science and technology studies, but also to the sociology of education, cultural studies and social research more broadly.

In this book I argue that the benefits of everyday science learning practices are only partially public. Participants' exclusion from everyday science learning practices echoed through the different contexts of their daily lives. Difficult experiences with science at school, home and at work echoed their perceptions of everyday science learning as exclusive. These experiences confirmed their expectations, making their non-participation a resilient feature of both their lives and the systems that excluded them.

Being unable to access everyday science learning opportunities can be considered a form of marginalisation and oppression. This is especially the case in societies where engagement with science can be considered key for cultural participation, political voice, education and entertainment, let alone issues of health, employment or well-being (Atwater, 2012; Harding, 2006; Orr & Baram-Tsabari, 2018; Orthia, 2013; Reardon & TallBear, 2012; Young, 1990). I argue that exclusion from everyday science learning is not a question of rebranding and changing perceptions of science or science on television, but instead goes to the core of how everyday science learning is understood and practiced.

Despite appeals for socially inclusive practice at different points in time, certainly across the museum sector, the research carried out for this book suggests many everyday science learning practices remain exclusive resources predominantly used by the more enfranchised groups of society (Association of Science and Technology Centres, 1987; Atkinson, Siddall, & Mason, 2014; Sandell, 1998, 2002; Sandell, Dodd, & Garland-Thomson, 2010). If we hope to develop inclusive, equitable everyday science learning practices that disrupt and transform processes of social reproduction and contribute to dismantling the structural inequalities embedded in our societies, we have to embrace change. To do this we need to better understand how exclusion operates and to respect the experiences and stories of those for whom this system does not work.

The research setting

The research for this book was carried out between the central London boroughs of Southwark (where I lived for the last ten years) and Lambeth. Specifically, my field sites were the neighbourhoods in these two boroughs strung like pearls along the route of the number 68 bus. Travelling south from the river Thames on the 68 bus you are more likely to hear Latin Americans speaking Spanish and Portuguese, Francophone Africans and Polish speakers than English. The 68 bus route took me south from my home to the Asian and Afro-Caribbean community groups and north to the Somali, Sierra Leonean and Latin American communities groups. Beyond the fantastically named neighbourhood of Elephant and Castle, the 68 bus carried me to the two universities where I first studied and later worked while this project unfolded.

Not for nothing is the 68 known locally as the slowest bus in London. The central artery of this study and the 68's route, the Walworth Road, is a busy, shop-lined street, filled with traffic. As Suzanne Hall (2012) has so meticulously documented, ethnic diversity and class mark Walworth, with Turkish and African supermarkets, Mixed Blessings, the Afro-Caribbean bakery, Figaro's the barber shop, and a smattering of fried chicken shops, nail parlours and international money senders. The social housing estates around the Walworth road were initially famous in the 1960s for heralding a new dawn for urban housing. But by 1997 these neighbourhoods were famous instead for the poverty experienced by their inhabitants. Inhabitants described as "forgotten people" in Tony Blair's first speech as Prime Minister, delivered on the Aylesbury Estate, a stone's throw from the Walworth Road (BBC, 1997; Skeggs, 2004).

Walworth, like the other neighbourhoods participants lived in, is frequently characterised in terms of disadvantage. During this study, for instance, the UK government positioned Walworth as an area of multiple-deprivation. This term was used to explain that the area suffered from a higher crime rate, higher unemployment, worse education infrastructure and more ill health (amongst other things) than other areas of London (Department for Communities and Local Government, 2011). In government terms, these are the features of inner city, multi-ethnic, working-class life.

It was important for this research that research participants lived somewhere with lots of everyday science learning opportunities on their doorstep which they were, nonetheless, not involved with. As such, neighbourhoods like Walworth in central London presented an ideal location for this research. Where someone lives in London reflects their social status through work and housing practices that separate rich from poor, majority from minority 'racial'/ethnic groups and those with legal status from 'illegal' status (Hall, 2012; Sassen, 2001; Sturgis, Brunton-Smith, Kuha, & Jackson, 2013).

While some residents of Southwark and Lambeth are wealthy, many are not. As a diverse, global city London is home to migrants from around the world and those who are not wealthy look for affordable housing in neighbourhoods like Walworth. But housing is hard to come by. During my fieldwork whole council estates[1] in Southwark were demolished to make way for new homes, as the "Now here", "Nowhere" graffiti on the Heygate Estate testified to prior to its demolition (see Figure 1.1). Eye-wateringly expensive blocks of new flats mushroomed along the 68 bus route, ousting estate residents as gentrification and profits fed into the crisis in affordable housing (Minton, 2017).

Figure 1.1 Now here/Nowhere graffiti on top of a block on the now demolished Heygate Estate, Elephant and Castle 2013

Saskia Sassen argued, "global cities are a key site for the incorporation of large numbers of immigrants in activities that service the strategic sectors. The mode of incorporation is one that renders these workers invisible" (2001, p. 322). Thus in cities such as London, not only can populations be considered "super-diverse" (Vertovec, 2007, p. 1025), but poverty frequently overlaps with 'race'/ethnicity for racialised groups as a result of the way employment is structured. Thus, while globalisation has brought new opportunities, it has exacerbated existing patterns of privilege and disadvantage (Massey, 1994). Nowhere are these patterns more evident than in cities like London, marked as Les Back has argued, by "the co-presence of multiculture and racism" (2007, p. 118). London's two biggest science museums, the Natural History Museum and the Science Museum seemed, in participants own words, "a long way from Walworth".

But Walworth is not a neighbourhood unaffected by science stories. It is marked at its northern tip by a small, 'blink-and-you'll-miss-it' sign commemorating the now demolished home of Michael Faraday, famous for his 19th-century studies of electromagnetism. The Walworth road continues south in an almost-straight line from Elephant and Castle to Camberwell. There it ends next to an image of the Camberwell Blue butterfly, first identified by Victorian lepidopterists in Camberwell, now immortalised in mosaic over the entrance of a local shopping centre. So although science was not described as an interest by most of the participants in this study, it could still be traced through Walworth and participants' lives.

The study: an overview

This book describes a qualitative study that followed an ethnographic approach, working in partnership with participants who rarely if ever took part in everyday science learning. Fifty-nine people from five grass-roots community groups[2] agreed to take part in the research and worked with me on and off for a fieldwork period of approximately two years. Five community groups from southeast London ultimately took part in the research I discuss in this book: an Afro-Caribbean group (n = 7), a Somali group (n = 6), a Sierra Leonean group (n = 16 and five children who came on the visit), a Latin American group (n = 17 and two children who came on the visit) and an Asian group (n = 13).[3] Although seven children came on the accompanied visits, I do not draw on their data, except indirectly when, for instance, their families mentioned them, as they were not themselves participants in the research. Participants ranged from 18 years old to 76 years old and across the 59 participants, 41 were female and 18 were male.

The research described here draws on data from five focus groups (one per community group), 30 interviews (including repeat interviews with some participants), four accompanied visits to everyday science learning activities chosen by the groups and around 1,000 pages of field notes and emails. Focus groups, interviews and accompanied visits were audio recorded, transcribed and anonymised during transcription, while field notes and emails were anonymised during analysis, with pseudonyms chosen by participants. The everyday science learning sites and institutions involved in the accompanied visits have also been anonymised in this book, and as a result, some details about certain exhibits or interactives have been changed. If you are interested, a more detailed description of the methods I used can be found in the Appendix.

Key terms and concepts

It is worth discussing key terms, concepts and language, firstly, in terms of everyday science learning, secondly, about people and, thirdly, to describe and think about publics.

What is everyday science learning?

Because I took an ethnographic approach to the study this book is based on, I used as broad as possible a definition of what the many everyday science practices that act as intermediaries between people and science in the public sphere might be. I wanted to make sure my own definitions did not unduly restrict what participants talked about or did during the two years that we explored equity, exclusion and public science together. Of course, this broad approach has the additional effect of making it hard to limit language and terminology to a single field when writing about the whole study.

Although I write about school, university and work to provide some context for participants' perspectives, this book sits mainly at the overlap of two jargon heavy areas of research and practice that combine surprisingly rarely: science communication and science education (Baram-Tsabari & Osborne, 2015; Dillon, 2011). Rather than taking a side, in this book I draw on ideas and language from both alongside the term 'everyday' to try to do justice to participants' experiences. In doing so I am all too aware that activities like museum visiting or political consultations are far from everyday, even for those groups who *do* regularly take part, while activities like watching television fit with the term somewhat better for many people (Scott, 2009). Thus, part of the background to this book is drawn from the sociology of practice and everyday life, the sociology of science in everyday life and pedagogic arguments about how learning happens across the different spaces and times of our lives (Bourdieu, 1990; Certeau, 1984; Lefebvre, 2008 [1968]; Lemke, 2000; Michael, 2006; Wortham, 2008). I draw on these ideas to think about everyday science learning practices as a mixture of different kinds of activities, in different spaces, with different motivations and actors. I explore how these practices are rooted (or not) in people's daily lives and their expectations about those practices.

The landscape of science communication is a shifting and fractured one, both in practice and research. It ranges from politically oriented activities (such as policy consultations) to those with cultural or educational motivations (such as museum visits or television programmes). Other terms are used to refer to science communication in this space, such as 'public understanding of science' and 'public engagement with science', and all have been contested over the years (Bauer, 2009; Davies, 2013; Miller, 2001; Smallman, 2016).

Much of the debate over terms within science communication has been about whether science is being communicated 'at' publics or 'with' publics (Davies & Horst, 2016; Lock, 2002; Medvecky & Leach, 2017; Wynne, 2006). However, most of the practices that participants discussed were firmly in the realm of communication 'at' publics and, as others have discussed, the shift from communication to engagement in the research literature may be less fully realised in practice than might be hoped (Burchell, 2007; Irwin, Jensen, & Jones, 2012). Thus in this book I use the term 'science communication', but define it broadly to include the full range of political and cultural/educational activities (see also Davies & Horst, 2016).

The neighbouring field of science education, while it overlaps in both practice and research with science communication, draws on different theories and, at times, different motivations. The sub-field of informal science learning is perhaps the closest part of education research to science communication. The use of the term 'informal' is meant to imply a difference in learning setting. Informal learning happens outside schools or universities, which, in turn, are referred to as 'formal' education. Informal science learning is used therefore to describe a range of quite different activities, participants, aims and practices. It can mean attending a science-themed community club, watching science documentaries, pursuing science-related hobbies or visiting museums, science centres, zoos, aquaria, botanic gardens or science festivals. It can also refer to activities based on engineering, technology or mathematics as well as those subjects captured by the term science.

As with science communication, the differences between practices and the language used is contested (Bell et al., 2009; Stocklmayer et al., 2010). Thus people also refer to 'out of school' and 'after-school' learning, life-long learning and 'designed' learning spaces[4] such as museums (Aguilar & Krasny, 2011; Bevan, 2017; Dawson, 2017; Falk & Dierking, 2012). For the purposes of this book I will use the term 'informal science learning' since it makes enough room conceptually for encounters with science on television, in books, newspapers or in conversations with friends, as well as in spaces like science museums.

Although it would make the book easier to write (and possibly easier to read), for me, eliding science communication with science education and informal science learning does not work. Certain practices in these two fields of research and practice are similar. But these fields are based on different research and practice traditions, each with a distinct history, as well as different theoretical and political commitments. As a result, relying only on the term 'science communication' minimises what, to me, is the most important, transformative part of this field of practice from a social justice perspective: the potential for learning (Freire & Freire, 1992; hooks, 1994). Equally, although I would argue learning and access to information are the cornerstones of participatory democracy and its allied public practices, I am loath to write only about informal science learning because this risks obscuring the more politically oriented public science activities participants discussed. Nonetheless, because of certain decisions made by participants during the project there is an emphasis on museums and science centres in parts of this book (see Chapter Six). Finally, the terminology of informal science learning invokes the language of education and learning, and can result in a tendency to focus on young people as participants and activities that are for, by or with young people. However, I doubt that many adults consider watching a nature documentary in the evening an 'out of school' experience. In turn, much of the science communication research focuses on adults, rather than young people. So for me neither term captures quite what I need it to.

The phrase 'everyday science learning' is my attempt to bridge some of these issues. I use everyday science learning in this book as a way to describe the multitude of practices that act as intermediaries between people and science in the public sphere, with their differing practices, practitioners and commitments. Throughout this book I use the term everyday science learning when talking in broad terms about participants' perspectives and experiences of public science activities, or as Ibrahim from the Sierra Leonean group put it, "the science experiences of your everyday life". Throughout the book therefore I use 'science communication' and 'informal science learning' when I am referring in particular to these areas of practice and research, 'everyday science learning' when I am referring to the broadest possible definition of how people engage with science and specific terms, such as television watching or science centre visiting, as appropriate. And, like Maureen Burns and Fabien Medvecky (2016), I suggest these practices can divide science and publics more than they bring them together.

Describing people

How we describe people is just as important as how we describe a field of research and practice. Inherent in social research on experiences of exclusion is the risk of essentialising those involved and contributing to problematic constructions of non-participants, non-visitors, the excluded, Others, or underprivileged groups. For instance, one pernicious implication of identifying and representing people or groups as Other – whether as non-participants or excluded – is the risk of framing certain groups as themselves at fault (Bell et al., 2009; Levitas, 2004).

In this book I use Dorothy Holland, Debra Skinner, William Lachiotte Jr., and Carole Cain's (2001) framing of identities as socio-culturally mediated constructs. These constructs, while fluid, are contextual and embodied in ways that matter. Holland et al. (2001, p. 271) describe facets of identity such as 'race'/ethnicity, gender, class, sexuality, (dis)ability as "social positions". Social positions are constructs in the sense that Judith Butler (1990) writes about gender as performative, and in the sense that Valerie Walkerdine (1990) describes class and gender as working together to support certain kinds of stories or fictions about people. As intersectional feminist research reminds us, however, not all identity performances or fictions are open to all bodies, in all spaces (Ahmed, 2017;

Mohanty, 2003; Paechter, 2007; Puwar, 2004). As such, social positions and how people are described in research matters and, necessarily, such terms change over time and space and come with their own particular allegiances.

In this book I have tried to describe, analyse and represent people and their specific experiences while being mindful of contributing to constructions of Others (Michael, 2012; Young, 2000). The terms we use and how we use them have implications. I use the term 'minoritised' to encompass the ways in which structural inequalities mapped onto the multiple and intersecting subjectivities of participants' lives. For me, this term, and those I discuss below, place the emphasis onto how social positions are social constructions, rather than practices and outcomes that fall somehow naturally and inevitably from who people are.

I use the language participants used to describe their 'race'/ethnicity and the terms they told me they preferred wherever possible (for instance, Black women, Latin American, Colombian, British-Asian, from Somaliland and so on). My participants were vocal about their critiques of terms like Black and Minority Ethnic (often shortened BAME or BME) or Ibero-American, terms used to group people for political reasons in ways they were not comfortable with. For instance, during the research period there were significant political struggles within the Latin American group not to be classified as Ibero-American in data about who lived in London, a subject they felt so strongly about that they organised more than one protest march about it.

When specific language is not appropriate I use the terms 'racialised groups' when writing about participants, depending on the emphasis needed. As you have already seen, I also use the combined term 'race'/ethnicity. Although they are sometimes used interchangeably, the concepts of race and ethnicity have complicated histories and different political and theoretical commitments (Gilroy, 2008; Hall, 1996; Hill Collins & Bilge, 2016). Nonetheless, building on Yasmin Gunaratnam's (2003) work, I use these terms in combination because, for participants, issues of cultural background, language, skin colour, religion, migration and region or country of 'origin' mixed together in how they saw themselves as racialised groups in London. Thus I have retained both words.

Similarly, I use the term socio-economic status when I am writing specifically about people's financial position, and the term class when I am writing about class in broader terms than economics alone (Levitas, 1998; Savage, 2010; Skeggs, 2004). I use the term gender in this book and, as above, refer to participants in gendered terms as they referred to themselves.[5] In using terms like these I acknowledge there is a delicate balance between recognising exclusion and attempting to understand the specifics of social reproduction on one hand and the potentially damaging use of labels on the other hand. The problems and politics of identification and representation are no small matter and, as Iris Young has argued, "all systems and institutions of representation group individuals according to some kind of principles, and none are innocent or neutral" (Young, 2000, p. 143).

This book is based on qualitative research; thus participants are not meant to represent everyone in their immediate community group, ethnic or socio-economic background, age, sexuality or gender, nor are the groups intended to represent everyone from their broader communities. Doubtless I will have changed my mind about some of this language in the future, but for now, this book is an attempt to provide concrete examples of attitudes towards and experiences of everyday science learning in an attempt to reimagine these practices through the lens of equity and social justice. As such in writing about all five groups throughout this book, I do not mean to suggest that their experiences were identical, but rather, to draw out how their experiences shared certain similarities in terms of equity, exclusion and everyday science learning. For instance, I argue throughout the book that their experiences were shaped by socio-historic contexts, not least imperialism, colonialism and structural inequalities that positioned them as minoritised people, and that these racialised and classed practices were embedded in the everyday science learning experiences they told me about.

What is public? Who are the public?

Finally, in thinking about people and their relationships with everyday science learning it is important to consider what is 'public' and who 'the public' are, particularly in terms of *who* can access that which is meant to be *public* (Benhabib, 2002). In this book, I take the position that 'public' means, as Young (1990) succinctly put it, "what is open and accessible" (p. 119). While this definition may seem simple, I found it a useful touchstone in this work. Which everyday science learning practices were open and accessible for participants? How were certain practices closed and inaccessible?

Young's definition of public also works with feminist critiques of ideas about the public sphere based on the work of Jürgen Habermas (1989). Feminist scholars concerned with gender, 'race'/ethnicity and class (amongst other issues) have criticised such models of the public sphere as fixed on dualistic distinctions between public and private, special and everyday, male and female (Ebrey, 2016; Fraser, 1990). In its simplicity Young's view of what is public supports a broad view of everyday science learning. Thus, rather than focusing only on special events outside the home, for example, activities that make science public in Young's sense could range from television watched in a bedroom, to taking part in a town hall meeting about local pollution, to visiting a museum.

Alongside the concept of public sits that of 'the public'. The notion of the public is, in some sense, a defining one for thinking about everyday science learning (Aikenhead, 2002; Bensaude-Vincent, 2001; Burns & Medvecky, 2016; Holliman, Whitelegg, Scanlon, Smidt, & Thomas, 2009). But that does not mean people agree on who the public are or how they should be understood. In this book I build, as others have done, on John Dewey's (1927) influential idea that publics are pulled into existence through practices, such as science policy consultations or visiting zoos (Borun, Chambers, & Cleghorn, 1996; Marres, 2007;

Michael, 2009). From this perspective, it is hard to imagine the public as a kind of ill-defined but fixed entity, lurking somewhere in the world.

Dewey's (1927) much used idea is useful because it helps me to think about how publics are brought into being through practices. In particular, and contrary to much of the research on science communication and informal science learning, I build on ideas from Howard Becker (1963), Nirmal Puwar (2004) and Sara Ahmed (2012) to think about how practices create publics both as 'insiders' and 'outsiders'. That is, rather than focusing only on those publics involved in a given practice, this book is an exploration of how practices create publics through exclusion and what these publics make of the situation.[6]

I also draw on an inclusive, participatory and democratic model of 'the public' from social justice theorists whose work has sought to value difference by understanding publics as heterogeneous and active in global, multi-cultural societies (Benhabib, 2002; Young, 2000). In response to social changes over the last 30 years the public have increasingly been framed as more heterogeneous and plural in research about everyday science learning. But as Jodi Dean (2002, p. 36) has argued, "adding an 's' to the theorisation of the public in no way suffices as a response to this challenge". Ideas from social justice can help to provide a framework for understanding what these practices of inclusion and exclusion mean. I make this point because as research shows us, not everyone gets to be included in 'the public' in the UK (Bhopal, 2018; Gilroy, 2002; Puwar, 2004). Assumptions about who is implied in terms such as 'the public' or 'citizens' are marked by 'race'/ethnicity, gender, class, ability/disability, sexuality and the other intersecting subjectivities of people's lives. Thus, it is helpful to keep in mind that publics are brought into being in the light *or shadow* of specific practices.

Pathways through the book

This book follows the format of a research monograph to tell the story of what happens when people from racialised groups encounter science through the multiple and various practices of everyday science learning. As alluded to in this chapter however, because the data generated with participants followed a broadly defined view of everyday science learning, this book covers a range of different settings – from visiting science centres and museums to watching television at home to participation in science-related policy events (admittedly briefly) – as well as different themes such as attitudes towards science, experiences of science at school and as a career, patterns of cultural science consumption and theories of exclusion and inclusion. In case you read books like me (index first and rarely from front to back), this part of the chapter maps out the content of each subsequent chapter, so you can find your own path through the book. In addition, each chapter closes with a summary of that chapter, to provide an alternative way to steer through the book.

Chapter Two is the first of two chapters in this book that are mainly theoretical (the second being Chapter Seven). Chapter Two sets out the key ways that exclusion and inclusion are currently framed in everyday science learning. That is, exclusion is all too often framed as the result of deficits on the part of the excluded, while inclusion is understood as a kind of assimilationist crusade. I then argue that we might be in a better position to understand exclusion and ultimately to work towards inclusive, equitable practices if we think more carefully about the roles of structural inequalities in everyday science learning. I discuss structural inequalities – racism, class discrimination, sexism and the many other forms of oppression experienced by people on the basis of their social positions – in terms of theories of social justice and ideas about social reproduction. I argue that we must move beyond deficit approaches to understanding exclusion and crusade approaches to understanding inclusion if we are to disrupt and transform everyday science learning and redevelop meaningfully inclusive practices.

Chapter Three is the first of five empirical chapters. Chapter Three takes a cultural consumption approach to understanding in broad terms participants' involvement in everyday science learning and, for comparison, other cultural and political practices they were more involved in. If you want to know more about patterns of participation in everyday science learning at the international scale, as well as what that looked like from participants' perspectives (who watches science on television, who visits museums), then this is the chapter for you. In taking a broad approach to understanding cultural consumption, I argue that the different fields involved in everyday science learning were marked by structural inequalities such that they operated as restricted fields of cultural production. In other words, not only were participants unable to access most of the everyday science learning practices we discussed, the practices they valued were rarely recognised or represented beyond their communities. The data discussed in this chapter suggest that focusing on people's *assets* rather than perceived deficits may present a better starting point for thinking about inclusion and equity in everyday science learning.

Chapter Four focuses on participants' attitudes towards and experiences of science. It starts by discussing participants' dispositions towards (or in this case, away from) science and argues these were shaped by structural inequalities. I then tell the stories of three participants – Mr Bhakta from the Asian group, Ibrahim from the Sierra Leonean group and Fatima from the Somali group – to show in more detail the effects of structural inequalities on people's lives and their relationships with science. These three participants stood out from their friends and families because they all liked science and tried to pursue it in their own ways, at school, university, as a career and as a hobby. As I discuss, however, in a system rigged against them, an interest in science was not enough to unlock science learning or science careers. I argue that we ought instead to understand participants' shared dispositions against science as structured by systematic inequalities that cut across their experiences of science in ways that are pernicious and enduring. The socio-political and historic roots of structural inequalities embedded across the different systems participants encountered in their lives, from education to employment, are not to be taken lightly.

Chapter Five explores participants' experiences and perceptions of exclusion from everyday science learning. The chapter starts by discussing exclusion from everyday science in broad terms, across practices from science on television to political consultations, and then focuses in on science museums as a specific example. In this chapter I argue participants experienced their exclusion from everyday science learning as forms of oppression, in particular, as cultural imperialism and powerlessness marked by racist practices that intersected with class discrimination and, for some, sexism. Participants' exclusion was embodied in the differences they felt between themselves and the somatic norm of white, middle- and upper-class bodies who used everyday science learning resources.

Practices of exclusion produced a visceral, embodied sense of alienation for participants that led them to reject everyday science learning at the same time as they were excluded. I argue therefore that exclusion from everyday science learning must *also* be understood as non-participation and suggest these operate as two sides of the same coin, creating a resilient exclusive system. I argue in Chapter Five that we must think carefully about what participants might need in order to stop rejecting everyday science learning. I suggest that 'quick fixes' and solutions designed around institutional interests to pick 'low hanging fruit' cannot and will not make a dent in how racism, class discrimination, sexism and their intersections structure everyday science learning.

Chapter Six discusses the accompanied visits carried out as part of the research for this book. Four of the five community groups chose to take part in an everyday science learning activity with me. Three groups chose to visit science museums and one group chose to visit a science centre. This is the chapter for you if you want to read about how exclusion was embedded in the practices of the science museums and the science centre we visited. The visits testified to how much extra work was required of participants to simply be in these spaces. Being inside the buildings did not mean participants were included. Instead, participants were Othered by the extent to which they were made to feel that they did not fit in. Thus, not only was accessing science learning opportunities made difficult, but participants felt all too aware that they were not represented, respected or particularly welcome in the science museums and science centre we visited. Drawing on this data, I argue in Chapter Six that we must think about how best to disrupt and transform everyday science learning practices to foster equity and inclusion as a matter of urgency.

Chapter Seven opens with empirical data and goes on to present a theoretical framework for how we might understand equity and inclusion in everyday science learning based on the data discussed and arguments made throughout the book. It starts by asking whether there is any point working with everyday science learning? The bulk of this book focuses on understanding exclusion so that we might have a better foothold from which to work towards inclusion. The book does, however, paint a bleak picture of everyday science learning. The empirical section of Chapter Seven draws on (the admittedly rare) moments from the accompanied visits where participants had more meaningful encounters with science or with the spaces we visited. Building on these encounters, I argue it is possible to disrupt and transform everyday science learning practices *if we choose to*. Furthermore, I argue that disruptive, transformative changes are both necessary and urgent.

The second, longer part of Chapter Seven presents an empirically led, theoretically informed framework for thinking through what equity and inclusion in everyday science learning might mean. If you are interested in understanding more about developing everyday science learning practices that are meaningfully inclusive and equitable, this is the chapter for you. In Chapter Seven I discuss a framework that uses three critical lenses – infrastructure access, literacies and community acceptance – arranged on a spectrum from weak to strong forms of social justice. I discuss these lenses drawing on data from the other chapters in this book and illustrate them with examples of practice where possible. The framework is, I hope, a tool to support the thinking, planning, evaluation and change needed to develop equitable and inclusive everyday science learning practices.

Chapter Eight is not the traditional conclusions and implications chapter of an academic monograph, mainly because much of this work has been done already in Chapter Seven. Instead, I use Chapter Eight as an Afterword to discuss some of the aspects of the research for this book that remain tricky for me. These are, firstly, the question of what we might do about science content. In my experience scientific content is often seen as sacrosanct, but here I wonder what would happen if it was not. Secondly, I reflect on some of the tensions that arose in thinking, talking and writing about racism during and after the research carried out for this book. While participants also discussed issues of class discrimination and sexism, these were always rooted in their experiences of racism. I have learnt, however, that one of the side effects of widespread institutional whiteness in everyday science learning is that people struggle to engage with racism. Thirdly, I discuss the implications for aspects of this book beyond everyday science learning. This chapter finishes with a discussion of some of the contemporary issues that preoccupy me when I think about everyday science learning and equity today, an attempt to take the temperature of the water surrounding this research.

Finally, at the very end of the book you can find an appendix that describes in more detail the research methods, participants, fieldwork sites and analytic approaches I used in the research carried out for this book.

It is with all these ideas in mind that I turn to trying to understand exclusion from everyday science learning in order that we might be better placed to understand and work towards inclusive and equitable practices.

Notes

1 Council estates are (or were) publicly owned social housing, but many homes on council estates are now privately owned in Southwark.
2 My initial conversations with community gatekeepers before I began recruitment for the study taught me the difference between grass-roots community groups and service-provision community groups. The latter are usually organised by an institution

(a funder, local or national government, education systems and so on) and provide a service such as English language courses or health care. Grass-roots community groups were, in contrast, usually organised by community members and provided opportunities for members to socialise rather than pursue particular services. It was these grass-roots community groups that I contacted for the study since I wanted members to be able to participate (or not) without any implications for their access to services.

3 Notably two groups in particular used broad terms (Asian and Latin American) to describe themselves, purposefully using what they considered to be more inclusive language, since their members came from a range of different countries.

4 At one point I briefly worked in science documentary production. When people distinguish between designed informal science learning spaces such as museums and other informal science learning opportunities that, by association, seem less designed, I am always reminded of how much work goes into designing television programmes.

5 It is also worth noting that while issues of sexuality, (dis)ability and other facets of social position around which discrimination can be experienced were certainly part of participants' lives, but these were not issues they discussed with me in relation to the research discussed in this book. As a result, while these issues are doubtless relevant to analyses of exclusion and inclusion, as attested to by other research (Cassidy, Lock, & Voss, 2016; Sandell et al., 2010), they did not appear as distinct themes in this project.

6 Special thanks to the attendees at the PCST Bellagio conference in 2017, organised by the indomitable Jenni Metcalfe. I hope you do not mind not being individually listed. Having the space to discuss how publics were constructed and how science communication could be understood with you all was beyond helpful and without doubt contributed to some of the subsequent framing of the arguments in this book.

References

Aguilar, O. M., & Krasny, M. E. (2011). Using the communities of practice framework to examine an after-school environmental education program for Hispanic youth. *Environmental Education Research, 17*(2), 217–233. doi:10.1080/13504 622.2010.531248

Ahmed, S. (2012). *On being included: Racism and diversity in institutional life.* Durham and London: Duke University Press.

Ahmed, S. (2017). *Living a feminist life.* Durham and London: Duke University Press.

Aikenhead, G. (2002). Science communication with the public: A cross-cultural event. In W.-M. Roth & J. Désautels (Eds.), *Science education as/for sociopolitical action* (pp. 151–166). New York: Peter Lang Publishing.

Association of Science and Technology Centres. (1987). *Natural partners: How science centres and community groups can team up to increase scientific literacy.* Washington, DC: Association of Science & Technology Centres.

Atkinson, R., Siddall, K., & Mason, C. (2014). *Experiments in engagement: Engaging with young people from disadvantaged backgrounds.* London: Wellcome Trust

Atwater, M. M. (2012). Significant science education research on multicultural science education, equity, and social justice. *Journal of Research in Science Teaching, 49*(1), O1–O5. doi:10.1002/tea.20453

Back, L. (2007). *The art of listening.* London and New York: Bloomsbury.

Ballard, H. L., Dixon, C. G. H., & Harris, E. M. (2017). Youth-focused citizen science: Examining the role of environmental science learning and agency for conservation. *Biological Conservation, 208*, 65–75. doi.org/10.1016/j.biocon.2016.05.024

Baram-Tsabari, A., & Osborne, J. (2015). Bridging science education and science communication research. *Journal of Research in Science Teaching, 52*(2), 135–144. doi:10.1002/tea.21202

Bauer, M. (2009). The evolution of public understanding of science: Discourse and comparative evidence. *Science Technology & Society, 14*(2), 221–240.

BBC. (1997). *Blair's speech: Single mothers won't be forced to take work.* Retrieved from www.bbc.co.uk/news/special/politics97/news/06/0602/blair.shtml

Becker, H. (1963). *Outsiders: Studies in the sociology of deviance.* New York: The Free Press.

Bell, P., Lewenstein, B., Shouse, A. W., & Feder, M. A. (2009). *Learning science in informal environments: People, places, and pursuits.* Washington, DC: The National Academies Press.

Benhabib, S. (2002). *The claims of culture: Equality and diversity in the global era.* Princeton, NJ and Oxford: Princeton University Press.

Bensaude-Vincent, B. (2001). A genealogy of the increasing gap between science and the public. *Public Understanding of Science, 10*(1), 99–113. doi:10.1088/0963-6625/10/1/307

Bevan, B. (2017). Out of school time. In K. A. Peppler (Ed.), *The SAGE encyclopedia of out of school learning* (pp. 539–564). Thousand Oaks, CA: Sage.

Bhopal, K. (2018). *White privilege: The myth of a post-racial society.* Bristol: Polity Press.

Bonney, R., Cooper, C. B., Dickinson, J., Kelling, S., Phillips, T., Rosenberg, K. V., & Shirk, J. (2009). Citizen science: A developing tool for expanding science knowledge and scientific literacy. *BioScience, 59*(11), 977–984. doi:10.1525/bio.2009.59.11.9

Borun, M., Chambers, M., & Cleghorn, A. (1996). Families are learning in science museums. *Curator: The Museum Journal, 39*(2), 123–138. doi.org/10.1111/j.2151-6952.1996.tb01084.x

Bourdieu, P. (1990). *The logic of practice* (R. Nice, Trans.). Stanford: Stanford University Press.

Bradu, C., Orquin, J., & Thøgersen, J. (2013). The mediated influence of a traceability label on consumer's willingness to buy the labelled product. *Journal of Business Ethics*, 1–13. doi.org/10.1007/s10551-013-1872-2

Burchill, K. (2007). *UK governmental public dialogue on science and technology, 1998–2007: Consistency, hybridity and boundary work. Paper presented to STEG group at King's College London.*

Burns, M., & Medvecky, F. (2016). The disengaged in science communication: How not to count audiences and publics. *Public Understanding of Science, Online First*, 1–13.

Butler, J. (1990). *Gender trouble: Feminism and the subversion of identity.* New York and London: Routledge.

Cassidy, A., Lock, S. J., & Voss, G. (2016). Sexual nature? (Re)presenting sexuality and science in the museum. *Science as Culture, 25*(2), 214–238. doi:10.1080/09505431.2015.1120284

Certeau, M. de. (1984). *The practice of everyday life* (S. F. Rendell, Trans.). Berkeley and Los Angeles: University of California Press.

Cho, S., Crenshaw, K. W., & McCall, L. (2013). Toward a field of intersectionality studies: Theory, applications, and praxis. *Signs: Journal of Women in Culture and Society, 38*(4), 785–810. doi:10.1086/669608

Davies, S. R. (2013). Constituting public engagement: Meanings and genealogies of PEST in two UK studies. *Science Communication, 35*(6), 687–707.

Davies, S. R., & Horst, M. (2016). *Science communication: Culture, identity and citizenship.* London: Palgrave Macmillan.

Dawson, E. (2017). Social justice and out-of-school science learning: Exploring equity in science television, science clubs and maker spaces. *Science Education, 101*(4), 539–547. doi:10.1002/sce.21288

Dean, J. (2002). *Publicity's secret: How technoculture capitalizes on democracy.* Ithaca, NY: Cornell University Press.

Department for Communities and Local Government. (2011). *The English indices of deprivation 2010.* London Department for Communities and Local Government.

Dewey, J. (1927). *The public and its problems: An essay in political inquiry.* New York: Holt.

Dillon, J. (2011). Science communication: A UK perspective. *International Journal of Science Education, Part B, 1*(1), 5–8. doi:10.1080/21548455.2011.552277

Ebrey, J. (2016). The mundane and insignificant, the ordinary and the extraordinary: Understanding everyday participation and theories of everyday life. *Cultural Trends, 25*(3), 158–168. doi.org/10.1080/09548963.2016.1204044

Eddo-Lodge, R. (2018). *Why I'm no longer talking to white people about race* (2nd ed.). London: Bloomsbury.

Falk, J., & Dierking, L. D. (2012). Lifelong learning for adults: The role of free-choice experiences. In B. Fraser, K. Tobin, & C. J. McRobbie (Eds.), *Second international handbook of science education* (pp. 1063–1080). London and New York: Springer.

Fraser, N. (1990). Rethinking the public sphere: A contribution to the critique of actually existing democracy. *Social Text, (25/26)*, 56–80.

Fraser, N. (2003). Social justice in the age of identity politics: Redistribution, recognition, and participation. In N. Fraser & A. Honneth (Eds.), *Redistribution or recognition? A political-philosophical exchange* (pp. 7–109). London and New York: Verso.

Freire, P., & Freire, A. M. A. (1992). *Pedagogy of hope: Reliving pedagogy of the oppressed.* New York: Continuum.

Gentleman, A. (2018, July 18). Revealed: Depth of home office failures on Windrush. *The Guardian,* p. 1. Retrieved from www.theguardian.com/uk-news/2018/jul/18/revealed-depth-of-home-office-failures-on-windrush

Gilroy, P. (2002). *There ain't no Black in the Union Jack* (2nd ed.). Abingdon: Routledge.

Gilroy, P. (2008). British cultural studies and the pitfalls of identity. In M. G. Durham & D. M. Kellner (Eds.), *Media and cultural studies* (pp. 381 395). Malden, MA and Oxford: Blackwell Publishing.

Gunaratnam, Y. (2003). *Researching race and ethnicity: Methods, knowledge and power.* London, Thousand Oaks, CA and New Delhi: Sage.

Habermas, J. (1989). *The structural transformation of the public sphere: An inquiry into a category of bourgeois society* (T. Burger & F. Lawrence, Trans.). Cambridge, MA: MIT Press.

Hall, S. (1996). Who needs "identity"? In S. Hall & P. Du Gay (Eds.), *Questions of cultural identity* (pp. 1–19). London, Thousand Oaks, CA and New Delhi: Sage.

Hall, S. (2012). *City, street and citizen: The measure of the ordinary.* London: Routledge.

Harding, S. (2006). *Science and social inequality: Feminist and postcolonial issues.* Urbana and Chicago: University of Illinois Press.

Hill Collins, P., & Bilge, S. (2016). *Intersectionality.* Cambridge: Polity Press.

Holland, D., Skinner, D., Lachiotte Jr., W., & Cain, C. (2001). *Identity and agency in cultural worlds.* Cambridge, MA and London: Harvard University Press.

Holliman, R., Whitelegg, L., Scanlon, E., Smidt, S., & Thomas, J. (2009). *Investigating science communication in the information age: Implications for public engagement and popular media.* Oxford and New York: Oxford University Press.

hooks, b. (1994). *Teaching to transgress: Education as the practice of freedom.* London and New York: Routledge.

Ipsos MORI. (2014). *Public attitudes to science 2014.* London Department for Business, Innovation and Skills.

Irwin, A., Jensen, T. E., & Jones, K. E. (2012). The good, the bad and the perfect: Criticizing engagement practice. *Social Studies of Science, 43*(1), 118–135. doi:10.1177/0306312712462461

Jasanoff, S. (2007). Bhopal's trials of knowledge and ignorance. *Isis, 98,* 344–350.

Lefebvre, H. (2008 [1968]). *Everyday life in the modern world* (S. Rabinovitch, Trans.). London: Allen Lane.

Lemke, J. L. (2000). Across the scales of time: Artifacts, activities, and meanings in ecosocial systems. *Mind, Culture and Activity, 7*(4), 273–290.

Levitas, R. (1998). *The inclusive society?* Basingstoke and New York: Palgrave Macmillan.

Levitas, R. (2004). Let's hear it for Humpty: Social exclusion, the third way and cultural capital. *Cultural Trends, 13*(2), 41–56.

Lewenstein, B. V. (2015). Identifying what matters: Science education, science communication, and democracy. *Journal of Research in Science Teaching, 52*(2), 253–262. doi:10.1002/tea.21201

Lock, S. J. (2002). The public understanding of science: A rhetorical invention. *Science, Technology & Human Values, 27*(1), 87–111.

Marres, N. (2007). The issues deserve more credit. *Social Studies of Science, 37*(5), 759–780. doi:10.1177/0306312706077367

Massey, D. (1994). *Space, place and gender.* Cambridge: Polity Press.

Medvecky, F., & Leach, J. (2017). The ethics of science communication. *Journal of Science Communication, 16*(4), 1–5.

Michael, M. (2006). *Technoscience and everyday life: The complex simplicities of the mundane.* Maidenhead and New York: Open University Press.

Michael, M. (2009). Publics performing publics: Of PiGs, PiPs and politics. *Public Understanding of Science, 18*(5), 617–631.

Michael, M. (2012). "What are we busy doing?": Engaging the idiot. *Science Technology & Human Values, 37*(5), 528–554.

Miller, S. (2001). Public understanding of science at the crossroads. *Public Understanding of Science, 10*(1), 115–120. doi:10.1088/0963-6625/10/1/308

Minton, A. (2017). *Big capital: Who is London for?* London: Penguin.

Mirza, H. S. (1992). *Young, female and black.* London and New York: Routledge.

Mohanty, C. (2003). *Feminism without borders*. Durham and London: Duke University Press.

Nelkin, D. (1995). *Selling science*. New York: W. H. Freeman and Company.

Orr, D., & Baram-Tsabari, A. (2018). Science and politics in the polio vaccination debate on Facebook: A mixed-methods approach to public engagement in a science-based dialogue. *Journal of Microbiology and Biology Education*, 19(1), 2–8.

Orthia, L. A. (2013). *Doctor Who and race*. Bristol: Intellect Books.

Osborne, J., & Dillon, J. (2007). Research on learning in informal contexts: Advancing the field? *International Journal of Science Education*, 29(12), 1441–1445.

Paechter, C. (2007). *Being boys, being girls: Learning masculinities and femininities*. Maidenhead: Open University Press.

Phipps, M. (2010). Research trends and findings from a decade (1997–2007) of research on informal science education and free-choice science learning. *Visitor Studies*, 13(1), 3–22.

Puwar, N. (2004). *Space invaders: Race, gender and bodies out of place*. Oxford and New York: Berg.

Reardon, J., & TallBear, K. (2012). "Your DNA is our history": Genomics, anthropology, and the construction of whiteness as property. *Current Anthropology*, 53(S5), S233–S245. doi:10.1086/662629

Sandell, R. (1998). Museums as agents of social inclusion. *Museum Management and Curatorship*, 17(4), 401–418.

Sandell, R. (2002). Museums and the combating of social inequality: Roles, responsibilities and resisitance. In R. Sandell (Ed.), *Museums, society, inequality* (pp. 3–23). London and New York: Routledge.

Sandell, R., Dodd, J., & Garland-Thomson, R. (2010). *Re-presenting disability: Activism and agency in the museum*. Abingdon and New York: Routledge.

Sassen, S. (2001). *The global city: New York, London, Tokyo* (2nd ed.). Princeton, NJ and Oxford: Princeton University Press.

Savage, M. (2010). *Identities and social change in Britain since 1940: The politics of method*. Oxford and New York: Oxford University Press.

Scott, S. (2009). *Making sense of everyday life*. Cambridge: Polity Press.

Skeggs, B. (2004). *Class, self, culture*. London and New York: Routledge.

Smallman, M. (2016). Public understanding of science in turbulent times III: Deficit to dialogue, champions to critics. *Public Understanding of Science*, 25(2), 186–197. doi:10.1177/0963662514549141

Stilgoe, J., Lock, S. J., & Wilsdon, J. (2014). Why should we promote public engagement with science? *Public Understanding of Science*, 23(1), 4–15.

Stocklmayer, S., Rennie, L., & Gilbert, J. K. (2010). The roles of the formal and informal sectors in the provision of effective science education. *Studies in Science Education*, 46, 1–44. doi.org/10.1080/03057260903562284

Sturgis, P., Brunton-Smith, I., Kuha, J., & Jackson, J. (2013). Ethnic diversity, segregation and the social cohesion of neighbourhoods in London. *Ethnic and Racial Studies*, 1–21. doi:10.1080/01419870.2013.831932

Vertovec, S. (2007). Super-diversity and its implications. *Ethnic and Racial Studies*, 30(6), 1024–1054.

Walkerdine, V. (1990). *Schoolgirl fictions*. London: Verso.

Wortham, S. (2008). Shifting identities in the classroom. In C. Caldas-Coulthard & R. Iedema (Eds.), *Identity trouble: Critical discourse and contested identities* (pp. 205–228). New York: Palgrave Macmillan.

Wynne, B. (2006). Public engagement as a means of restoring public trust in science: Hitting the notes, but missing the music? *Community Genetics, 9*(3), 211–220. doi:10.1159/000092659

Young, I. M. (1990). *Justice and the politics of difference.* Princeton, NJ: Princeton University Press.

Young, I. M. (2000). *Inclusion and democracy.* Oxford and New York: Oxford University Press.

Chapter 2

Understanding exclusion

In this chapter I focus on how exclusion, inclusion and equity are understood in everyday science learning. I start by briefly exploring community ticket schemes in science centres and what we might learn from them about how equity and exclusion are currently understood by those working in everyday science learning. In the second section I discuss misconceptions about how exclusion from everyday science learning works and in the third I discuss misconceptions about how inclusion works. In the fourth section I lay out the theories I use throughout this book as an alternative way of thinking about equity, exclusion and everyday science learning. I argue that building on ideas about social justice, structural inequalities and social reproduction gives us a more useful way to understand exclusion and, as I argue throughout the book, provide a more helpful way, ultimately, to think about inclusion and equity.

Practice challenges: having a golden ticket

I first heard the term 'a golden ticket' in the context of everyday science learning and inclusion/exclusion when I was talking to colleagues from a science centre. We had been talking about their efforts to develop a system to encourage more visitors to their science centre from communities who did not usually visit. For many science centres and science museums around the world getting inside the front door requires an expensive ticket. In exploring how issues of inclusion, exclusion and equity play out in science communication and informal science learning I found many institutions had community-based schemes to provide tickets for people who would otherwise be unable to afford to pay their entry fee. Practices like these can tell us a lot about how exclusion and inclusion are understood by the people involved in everyday science learning and raise interesting questions from an equity perspective.

In some places, such as the Monterey Bay Aquarium in California (US), community tickets are available through libraries in certain communities. In other places, such as Dundee Science Centre in Scotland (UK), community tickets are available in particular schools. In many ways these schemes are useful practices that remove what many see as the chief impediment to participating in certain forms of everyday science learning: the expense. Indeed, entry costs to science centres, science museums, science festivals and other place-based science communication and informal science learning practices are regularly described as the key barrier to participation (Ipsos MORI, 2003; Newman, McLean, & Urquhart, 2005). As such, their temporary removal for specific communities, when the lottery of ticket availability allows, is an interesting practice.

The name of these golden ticket schemes is quite revealing about the assumptions that underlie contemporary inclusive practice in everyday science learning. Let us think for a moment about 'golden ticket' language. Today, having a golden ticket is a colloquial phrase that refers to a device that ensures your success. Perhaps you have read Roald Dahl's (1964) novel, *Charlie and the chocolate factory,* where the phrase came from? First published in the 1960s in the UK, the book tells the story of Charlie, a boy whose family live in poverty, and four other children who find golden tickets hidden in bars of chocolate.

The tickets allow the children to visit the amazing but closed world of Willy Wonka's chocolate factory. Ultimately, Charlie's life is transformed by his golden ticket experiences when he and his family are invited to join Willy Wonka in his magical, chocolate-filled world permanently. But let's not forget what happens to the other children, the ones who did not or could not follow Willy Wonka's rules. In the book they come to a somewhat sticky end.

Of course, not all community tickets are called golden tickets. But in the course of three minutes today, in March 2018, I was able to find six examples of science centres and museums with golden ticket schemes via Google across three different countries, which suggests the idea is fairly widespread. Furthermore, a report titled "Effectively engaging under-represented groups", published in the UK by the Association of Science and Discovery Centres (ASDC), recommends using "golden ticket" schemes as a form of best practice for inclusion and equity in science centres (2014, p. 12). Notably, the report also recommends " 'Golden Ticket' schemes and community days" because science centres and museums experience seasonal peaks and troughs in visitation. Thus community ticket schemes allow "your organisation to spread the bookings across the year and welcome these visitors at times when you have the staff resources to maximise their experience" (Association of Science and Discovery Centres, 2014, p. 12). While this does seem at first glance like a good way to ensure such visitors benefit from a high staff-to-visitors ratio, a more cynical reading of this advice is that community tickets can be used to bolster low visitor numbers during off-peak moments.

To me, the language of golden tickets for science centre or museum visits and the story they draw upon speak to an assumption by some working within everyday science learning that these practices, venues and their content are wonderful, and if minoritised communities only know this secret, they would flock to them. This approach to equity and exclusion is fundamentally misconceived.

This kind of golden ticket approach to supporting minoritised communities to engage more with everyday science learning is flawed from an equity perspective.

As a practice, it focuses on keeping science and institutional practices the same, while the locus of change is with communities. A golden ticket approach focuses only on entry rather than on more substantial changes to practices such as hiring, content development, representation or even the associated question of the entry fee structure itself. In other words, golden ticket systems leave exclusive practice in everyday science learning unquestioned, thereby maintaining the status quo. Such a practice, I argue, is simply unable to disrupt or transform existing patterns of participation in everyday science learning. Instead, such practices give the appearance of inclusion, while maintaining the status quo and/or making things worse.

This story about one kind of practice designed to support inclusion and equity in science centres and museums is, for me, indicative of some of the troubled ways that equitable practice can be configured – that is, premised on problematic ideas about exclusion and, as a result, limited assumptions about how inclusion works. Ideological views about exclusion and inclusion can lead to practices that appear to be addressing equity, whilst actually reproducing inequalities. As Ahmed (2012) argues, much of the work done in the name of diversity and inclusion can operate to maintain systems of discrimination. In the example of golden ticket schemes, although they work on part of the question of access (by removing entry costs), they stop short of the more fundamental changes that I argue in this book are needed to make everyday science learning inclusive, in terms of producers, users and content involved in such practices. It is important to question therefore how meaningful such inclusion practices can be and what vision of exclusion and inclusion they rely on. A better understanding of how exclusion operates and how it is maintained and reproduced through everyday science learning practices is a useful step to rethinking equity and inclusion.

Empirical challenges: looking outside the box

A significant challenge in trying to understand why and how patterns of exclusion exist and persist in everyday science learning practices is the limited amount of empirical research on the subject. This is not to say that we don't know a huge amount about everyday science learning! Rather, research on specific everyday science learning practices such as museum visiting, citizen science activities, television watching or policy consultations has tended to focus on those who already participate and what benefits their participation may provide for them.

While such research provides valuable insights about existing participants, the risk of inward-facing research is that it may, quite literally, be unable to look outside the box. In other words, the majority of everyday science learning research is constrained because of a tendency to "consider only those who are in the system at a given moment, excluding those who have been excluded from it" (Bourdieu & Passeron, 1990, p. 159). Thus, while there is a great deal of empirical research on, for instance, how visitors to museums, science centres, aquaria or botanic gardens behave, learn, socialise or recall their visits (see for example

Falk & Storksdieck, 2010; Packer, 2008; Tunnicliffe, 2008), there is comparatively little research on questions of access, inclusion/exclusion or the attitudes and experiences of those outside these systems.

Of equal concern is the narrow framing of the available research about exclusion. The most comprehensive mapping of research on everyday science learning participation available to date comes from the world of informal science learning (that is, practices in museums, science festivals, planetaria and similar settings), in three studies all of which systematically reviewed the literature (Bell, Lewenstein, Shouse, & Feder, 2009; Falk et al., 2012; Falk et al., 2015; Fenichel & Schweingruber, 2010).

The Falk et al. (2012/2015) research review is revealing in terms of mapping research focuses and gaps since it reviewed the international academic literature, while the other two reviews concentrated on the US context. Of the 553 peer-reviewed, academic articles reviewed by Falk et al. (2012), only 27 addressed participants, visitors or audiences who could be considered disenfranchised in some way; 8 focused on minority ethnic students, 10 on female students, 3 on minority ethnic families, 2 on female, minority ethnic students, 2 on minority ethnic groups, 1 on low-income students and 1 on low-income families. Over a 31-year period of internationally sampled, albeit English language papers, this is a surprisingly low number, especially given that concerns about equity in museums, cultural participation and science date back at least 30 years (Association of Science and Technology Centres, 1987; Gurian, 2006). We can see from these patterns that issues of equity and exclusion in everyday science learning are both narrowly framed and under-researched. Unfortunately, as I discuss in the following two sections of this chapter, the theoretical work on equity, exclusion and everyday science learning is just as limited as the empirical work and presents serious challenges to developing more meaningfully inclusive approaches.

Theoretical challenges: moving away from deficits and crusades

(Mis)Understanding exclusion through deficits

Understanding how people are excluded from everyday science learning has a troubled (racist/sexist/classed) history in the two fields of practice and research this book is most concerned with, science communication and informal science learning. As discussed in Chapter One, while ideas about the public are central to the concerns of researchers, practitioners and policy makers involved in everyday science learning, these concerns play out in multiple ways and rarely take structural inequalities and socio-historic context into account. In this section I argue that minoritised publics have been problematically misunderstood in relation to everyday science learning and, as a result, so too has their exclusion from it.

Studies of science and society relationships have long debated questions of deficits in knowledge and attitudes around science communication. But even the more nuanced models of public knowledge, attitudes and behaviours towards science and everyday science learning in practice and research often overlook the roles played by structural inequalities and the intersections of 'race'/ethnicity, gender, class and other social positions (see for example Sturgis & Allum, 2004). Although science communication practices have been described as "sharply unevenly distributed" (Rommetveit & Wynne, 2017, p. 134), questions of social justice, equity and exclusion are rarely discussed.

Informal science learning research is not much more helpful. Unlike science education in schools and universities, questions of inclusion and exclusion, access and equity have been under-researched and under-theorised. Perhaps unsurprisingly, as discussed in the previous section, research on informal science learning has instead typically focused on those who do participate and what benefits their participation may provide for them (Falk, 2009; Rennie & Williams, 2006). This limitation is particularly problematic since, as discussed in more detail in Chapter Three, descriptive survey data suggest that, at present, patterns of participation in science communication and informal science learning are socially uneven. As a result, research on informal science learning has concentrated around particular social groups, namely, the more enfranchised groups of a given society, those from the dominant ethnic groups, wealthier socio-economic backgrounds, living in urban areas. Thus research on the benefits of science museum visits may not be generalisable across diverse populations, nor offer clues about how to broaden the appeal of informal science learning institutions.

That such patterns exist in the literature is important. Social research methods contribute to the making of publics, how they are imagined and how their practices are understood (Savage, 2010). Social research is not neutral because, as with any research, values are involved, whether implicit or explicit. Publics can be researched as markets, as self-researching subjects or in ways that re-inscribe racist ideologies (Ebrey, 2016; Fourcade & Healy, 2016; Gillborn, 2010; Gillborn, Warmington & Demack, 2018).

Publics, and their participation, have been constructed through research practices in relation to science and everyday science learning (Michael, 2012). Indeed, the concept of an "imagined public" has become part of the public understanding of science lexicon (Marris, 2014, p. 90; Rommetveit & Wynne, 2017, p. 133). However, excluded or non-participating publics have remained either largely unexamined in such research or, more damagingly, have been imagined in negative (racist, classed, sexist) terms.

Where can we find publics who do not participate in science communication or informal science learning activities in the research literature? Such publics can be found in some of the large-scale surveys about cultural consumption, such as the UK Taking Part survey, and some of the more specific surveys about attitudes to science (Department for Culture Media and Sport, 2016; Ipsos MORI, 2016). Analyses of these surveys often identify publics through their non-participation in activities that involve science communication or informal science learning, albeit in problematic ways.

Take, for example, the most recent of the UK Public Attitudes to Science (PAS) survey reports (Ipsos MORI, 2014). The survey analysis demonstrates how excluded publics can be constructed in negative terms through perceived deficits and the kinds of values implicit in research design. A segmentation analysis identified two groups of respondents that we could consider as non-participants in or excluded from everyday science learning. The segments (the "concerned" and "disengaged sceptics") described people who felt ill informed about science, did not trust science or scientific regulation and rarely participated in everyday science learning activities (Ipsos MORI, 2014, p. 134). Notably, these segments included comparatively higher proportions of people from racialised groups, socio-economically disadvantaged backgrounds and women than the other segments. Building on research by David Gillborn et al. (2018) and Mark Taylor (2016), I suggest we ought to be cynical about how racialised, classed and gendered inequalities are embedded within survey design and analysis (not to mention qualitative research) in ways that valorise the practices and knowledges of dominant groups, but construct minoritised groups as deficient. While surveys like this one follow in the tradition of earlier science communication scholarship that sprang from concerns about public disinterest in science, the PAS surveys reveal troubling assumptions about the kinds of deficits that continue to shape how publics and their practices are understood.

The "deficit model" of science communication has been criticised for a long time (Lock, 2011, p. 17; Wynne, 1992). This model has been criticised for positioning publics as having the wrong (negative) attitudes to science and being unknowledgeable to boot (Bell, Davies, & Mellor, 2008; Davies & Horst, 2016). Thus much of the science communication work (policy, practice and research) in the UK from the 1980s onwards has been characterised by the scientistic assumption that, as Jon Turney (1998, p. 1) put it, "to know science is to love it". Not liking science and not wanting to participate in everyday science learning became, on the basis of these assumptions, almost unthinkable and something that those who know better ought to work to change.

The elitist values driving assumptions about deficits and non-participation in everyday science learning follow racialised, classed and gendered patterns about what dominant groups value in society (Bhopal, 2018; Bourdieu, 1984; Miles & Gibson, 2016; Schiebinger, 2007). For Kalwant Bhopal (2018), for example, framing people from racialised groups who do not follow dominant behaviours or attitudes as themselves problematic, somehow deficient and responsible for their own exclusion is a significant feature of racist societies. In terms of class, re-analysing data from surveys of cultural consumption in the UK, Mark Taylor (2016) found that elitist, classed assumptions about dominant cultural activities rendered working-class cultural practices invisible through implicit assumptions about which forms of culture mattered. As a result, certain groups were configured as without culture and thereby deficient. Similarly, in a meta-analysis of 30 years of UK programmes designed to encourage more women into science Alison Phipps (2008) found programme after programme sought to change the (wrong) attitudes and behaviours of women, leaving science, science education and science careers structurally unchanged. Thus, mistaken ideas about deficiencies on the part of groups who experience racialised, classed or gendered inequalities (and their intersections) all too often implicitly or explicitly structure the terms of research and practice. In other words, the people who experience the problem, become the problem.

If we think about the PAS segments alongside research about patterns of participation in everyday science learning practices discussed previously, we can see that socially dominant groups participate more (Ipsos MORI, 2014). The "concerned" and "disengaged sceptics" segments included higher proportions of people from backgrounds that are not socially dominant (Ipsos MORI, 2014, p. 134). Understanding their exclusion from and non-participation in everyday science learning might not be as straightforward as people in these two segments having the wrong sorts of attitudes and doing the wrong sorts of behaviours when it comes to science. If we think about the people in these two segments from an equity perspective, drawing on research about structural inequalities and their intersections, we might think instead about their access to science and everyday science learning, and their experiences of discrimination.

Structural inequalities – injustices that result from both people's unquestioned biases and oppressive features of political, cultural, educational or market forces – disadvantage certain groups in ways that can overlap or intersect (Crenshaw, 1991; Young, 1990). Without taking structural inequalities into account, ideas about exclusion and non-participation – whether in culture or politics, science or arts – often imply, as Ruth Levitas has argued, that excluded people "have the wrong values and attitudes" (2004, p. 49). Thus, analyses like the PAS one described above can frame the 'concerned' and the 'disengaged sceptics' as the problem, rather than examining whether everyday science learning practices are exclusive. As a result, publics that do not or cannot participate in science communication and/or informal science learning are imagined through such research as deficient and responsible, at least in part, for their own exclusion. If we are to address equity and inclusion in everyday science learning, we first need to move away from partial, deficit-oriented ways of framing exclusion and to think more rigorously about how such exclusion manifests and is reproduced for racialised and other minoritised groups.

(Mis)Understanding inclusion as a crusade

Combining ideas about social justice and publics for everyday science learning is not straightforward, especially given the problematic ways in which exclusion has been configured. In this section I argue that just as exclusion has been misunderstood in everyday science learning, so too has inclusion. In both exclusion and inclusion, a key problem about whose practices and knowledges are valued remains a central concern. Thus, as Levitas (1998, 2004) has argued, inclusion agendas are renowned for reifying dominant practices and values with little regard for the needs, interests or practices of marginalised groups. When it comes to inclusion/exclusion in everyday science learning a two-pronged evangelical approach seems to operate as a kind of crusade, one that reifies both the medium of everyday science learning and its content – science – in ways that protect the privilege of dominant groups.

First, dominant forms of everyday science learning practice (museum visiting, participating in government consultations) are reified and framed as worthwhile activities for members of the public to take part in for their own good, whether in terms of what people might learn or their duties as citizens (Levitas, 2004; Tlili & Dawson, 2010). Access to education and political voice are clearly important. But these are not neutral practices. It is crucial to remember that dominant cultural and educational practices (whether arts or science based) have long been criticised as a form of "moral regulation" from the perspectives of minoritised groups (McGuigan, 1996, p. 16). That is, people are expected to participate for their own good and the good of society, whether they want to or not (Belfiore, 2009).

In the UK, the New Labour government's social inclusion policies in the 1990s moved the emphasis of social inclusion away from the economics of poverty, participation in the labour market and social mobility, to focus instead on cultural participation, fundamentally shifting how social inclusion was addressed in practical terms (Levitas, 2004; Sandell, 1998). This shift was problematic and had lasting effects. Reframing economic differences as cultural differences obscures how poverty and wealth are reproduced in our societies. It moves government responsibility for those living in poverty away from financial support and welfare, and towards increased funding for cultural 'inclusion' activities. As Levitas (2004) has argued, we ought to be wary of attempts to gloss over the significant economic inequalities in our societies with the language of cultural inclusion. Nonetheless, we should still expect cultural, educational and political practices and institutions to be inclusive, but as a basic form of good practice, not as a special project that is somehow different from day-to-day practice (Golding, 2009; Lynch, 2011; Sandell, 2007).

Second, as Joan Solomon noted, science has been uncritically framed in culture, education and politics as "especially good for you" (2012, p. xiii). Privileging scientific information above other knowledges, by attempting to replace other perspectives or practices with those of science, has been described as scientistic and assimilationist (Lee, 1999; Lee & Buxton, 2010; Ogbu, 1992; Stanley & Brickhouse, 2001; Yosso, 2005). This evangelical, crusade approach assumes that exposing more people to science communication and informal science learning is de facto a good thing, as though science is a vitamin we all need more of.

Of course, from one perspective this approach makes sense. Science plays a powerful role in our societies. It is important to make sure everyone can access those knowledges, communities, practices and applications, to participate in them and benefit from them. But, as Londa Schiebinger (2007) has argued, we have to be careful about the project of Western science that we are trying to include people within. From an equity point of view, science has a socio-political history – as any human enterprise does – and it is marked by colonialism, racism, misogyny, ableism, homophobia and heteronormativity and other forms of oppression (see for example Cassidy, Lock, & Voss, 2016; Harding, 2006; Medin & Bang, 2014; Pollock & Subramaniam, 2016).[1] Thus, while science has many benefits, an uncritical, assimilationist perspective leads to a crusade approach because it overlooks the potential damage involved if we do not include transforming science itself as a key part of the project of transforming everyday science learning.

In combination, the two perspectives I have discussed here – framing exclusion as the result of perceived deficits on the part of the excluded and framing inclusion as an assimilationist kind of crusade – work together to absolve the state, institutions and individuals from responsibility for exclusion, while neatly sliding blame onto people who do not or cannot participate. These approaches to inclusion/exclusion, as discussed in the previous section, draw not only on racialised ideologies about white privilege, but classed and gendered power dynamics about whose cultures, knowledges and practices count (Ahmed, 2012; Bhopal, 2018; Gilroy, 2002; Levitas, 1998; Schiebinger, 2007; Yosso, 2005). Peggy McIntosh (1989), for instance, argued that, in terms of 'race'/ethnicity, these ways of thinking were pervasive and worked to reproduce white privilege. She wrote "whites are taught to think of their lives as morally neutral, normative and average, and also ideal, so that when we work to benefit others, this is seen as work which will allow 'them' to become more like 'us'" (McIntosh, 1989, p. 31). This perspective is similar to what Puwar (2004, p. 23) described as a "rescue" approach, where people could be saved by dominant groups (and in being saved, be remade in the image of their saviours). Thus, excluded and non-participating groups who are more likely to come from racialised groups and socio-economically disadvantaged backgrounds (as discussed further in Chapter Three), are not necessarily advantaged by crusade approaches to inclusion; in fact, quite the opposite.

If we turn to everyday science learning practice, reviews carried out in the UK and the US suggest that, with few exceptions, attempts at inclusive practice can be understood as assimilationist and crusade-like in approach (Bell et al., 2009; Falk et al., 2015; Fenichel & Schweingruber, 2010). For instance, tokenistic efforts to increase inclusion in an everyday science learning practice (such as science museum visiting) without considering the cultural, social, linguistic, political or other substantial changes that may be required for that practice to be appealing, accessible and equitable were found to be common (Bell et al., 2009; Fenichel & Schweingruber, 2010). Some golden ticket schemes discussed earlier could also be understood to fall into this pattern. Another example that springs to my mind was when someone from a science festival once explained to me, they were interested in inclusion in terms of the "low-hanging fruit". What quick, cheap changes could they make to improve inclusion without upsetting practice as normal? From a crusade perspective, it is non-participants, rather than dominant practices, who are expected to change.

Research with everyday science learning practitioners in museums and similar institutions across the cultural sector found that most people agreed that more inclusive, equitable practices needed to be developed (Feinstein & Meshoulam, 2014; Taylor & O'Brien, 2017; Tlili, 2008). Competing and contradictory rationales for inclusion were described however. Equitable aims were often lost alongside overriding institutional priorities about science content, subsumed into tokenistic aspects of reporting to funders or seen to be at odds with keeping existing visitors happy (Feinstein & Meshoulam, 2014; Tlili, 2008). Inclusion, although seen as important, was often of secondary importance to competing operational concerns.

Ahmed (2012) has argued that inclusion agendas can be framed in ways that protect established, institutional ways of being against the kinds of transformative changes that may be required to develop meaningfully inclusive practices. This kind of approach to inclusion draws on what Nancy Leong (2013, p. 2154) terms "racial capitalism". Racial capitalism reduces inclusion and equity to tick-box exercises that benefit primarily white institutions through seeming to become more diverse and inclusive, while continuing to exclude people from racialised groups. Crusade approaches to equity and inclusion can be understood therefore as 'inclusion-lite' and, more cynically, as strategically useful for everyday science learning institutions and practices unwilling to change.

We need to take seriously how structural inequalities operate in everyday science learning and reject views of inclusion/exclusion that explicitly or implicitly configure exclusion as the fault of the excluded. Crusade approaches to inclusion in everyday science learning that privilege dominant knowledges, practices and attitudes while pathologising others cannot support the development of meaningfully inclusive practice. In fact they prevent such changes from taking place. How then can we think about equity and inclusion in ways that go beyond assuming the excluded are deficient and simply require assimilation into science and science learning?

Understanding exclusion

This book is about what happens when people from minoritised backgrounds meet the messy world of everyday science learning practices that try to mediate relationships between science and society. Such practices and the research associated with them make science public and in doing so, make publics for science (Dawson, 2018; Dewey, 1927; Michael, 2012). To think about equity and inclusion in everyday science learning in ways that move away from assumptions about deficits and crusades I draw on a mixture of sociological theories, which I describe in this section. Building on theories of social justice, particularly those developed by feminist political philosophers Iris Young and Nancy Fraser, I frame structural inequalities – racism, class discrimination, sexism and other forms of oppression – as intersectional, drawing in particular on the work of Patricia Hill Collins and Sirma Bilge, as well as Kimberlé Crenshaw. I combine these ideas with Pierre Bourdieu's theoretical tools about social reproduction (that is, how social inequalities are reproduced so that the rich stay rich and the poor stay poor) to understand how exclusion happens in practice. This combination of theoretical tools forms the conceptual backbone of this book, informs the empirical analyses that I discuss and underpins the inclusion framework that draws the book to a close.

Structural inequalities and social justice

Blaming people for their own exclusion, and assuming inclusion is a simple matter of changing non-participants' attitudes and behaviours does not and cannot disrupt patterns of participation in everyday science learning. Instead, in this section I argue a more useful approach can be found by thinking about structural inequalities and social justice.

Understanding how disadvantages are reproduced in relation to everyday science learning is not as straightforward as looking for a person or organisation to blame. Rather, structural inequalities (overt and covert forms of discrimination and bias such as sexist assumptions about aptitudes for science) are embedded across institutions, policies and practices, as well as our everyday behaviours, in ways that maintain or exacerbate social advantages and disadvantages (Fraser, 2003; Young, 1990). As Young argued, inequitable practices can be understood as oppressive:

> oppression also refers to systematic constraints on groups that are not necessarily the result of the intentions of a tyrant. Oppression in this sense is structural, rather than the result of a few people's choices or policies. Its causes are embedded in unquestioned norms, habits, and symbols, in the assumptions underlying institutional rules and the collective consequences of following those rules.
>
> (Young, 1990, p. 41)

In this book I build on Young's (1990, 2000) view of exclusion as a form of oppression and the social justice concepts that go with it. Thus, exclusion is described here as a complex practice, enmeshed in the ingrained values, systems and behaviours of practitioners, policies and publics, as well as wider society. Here the idea of the everyday becomes important again. Exclusion from everyday science learning practices is, in this sense, systematic and complex, albeit sometimes hard to pinpoint since it is rooted in the mundane activities of everyday life (Bourdieu, 1984; Bourdieu & Johnson, 1993; Bourdieu & Passeron, 1990; Scott, 2009; Young, 1990, 2000).

Structural inequalities are practices of systematic discrimination against certain people by virtue of who they are (Hill Collins & Bilge, 2016). In this sense they are ideological practices (Gilroy, 2002). As a result, certain people, because of their skin, size, form, practices or manner are forced to contend with problems that other people never encounter. Though practices of discrimination shift over time and location, they are rooted in socio-political histories about who deserves privilege and respect and who does not. The structural inequalities I explore most in this book are racism, class discrimination, sexism and, importantly, their intersections.

In the research presented in this book, experiences of racism were the most significant structural inequality participants faced. Bhopal (2018, p. 5) describes racism as practices that "reinforce the position of whites at the expense of disadvantaging those from black and minority ethnic backgrounds". This definition is helpful because it emphasises the structural and systematic nature of racism (and in turn, provides a way to think about other structural inequalities, such as class discrimination and sexism as similarly systematic). As McIntosh (1989, p. 12) has argued, "I was taught to recognize racism only in individual acts of meanness by members of my group, never in invisible systems conferring unsought racial dominance on my group from birth". Thus to recognise racism we must think in structural terms about explicit practices and policies that disadvantage people from racialised backgrounds, as well as taken-for-granted institutional norms and implicit biases (Bhopal, 2018; Young, 1990). It is important, however, not to use a focus on structural practices and implicit biases to let ourselves off the hook. As Paul Gilroy (2002) reminds us, racism is reproduced through the work of individuals as well as institutions.

As discussed in Chapter One, participants in this book were women and men, across a wide age range, from racialised groups, who, at the time of the research were socio-economically disadvantaged. As a result, pulling themes of 'race'/ethnicity, class, gender or age neatly apart as though they might be stacked and restacked analytically belies the interwoven nature of people's lives and the structural inequalities that shape our societies. Thus, as Leah Bassel and Akwugo Emejulu (2017) argue, we must insist on thinking about these issues together. Although I have emphasised issues of racism in this section to provide an example of a particular form of structural inequality, we must think about structural inequalities not just as single facets of analysis, but in terms of how they intersect.

Taking an intersectional approach to understanding structural inequalities requires thinking about how for certain people, discrimination is experienced in multiple ways and may not be usefully reduced to one or other dimension of inequality or self (Crenshaw, 1991; Hill Collins & Bilge, 2016). Drawing again on the work of Hill Collins and Bilge (2016), I use their perspectives to understand an intersectional approach as one concerned with intersecting structural disadvantages, rather than individual identities. For example, to understand a Somali woman's experiences of everyday science learning, we must think, at least, about the structural inequalities that emerge at the intersections of 'race'/ethnicity, socio-economic position and gender. As such, I find the idea of intersectional approaches as "kaleidoscopic" useful, because it helps to imagine how structural inequalities are enmeshed, but cannot be reduced to one or another (Gunaratnam, 2015; Puwar, 2004, p. 10). Rather, each piece of the puzzle is refracted alongside the others, in different configurations, whichever way you look. Thus, throughout this book I try to understand participants' experiences of exclusion and everyday science learning in terms of intersecting structural inequalities and with reference to ideas about social justice.

Social reproduction and Bourdieu

In this section (and throughout the book) I build on Bourdieu's extensive work on social reproduction alongside ideas about social justice, to think through how structural inequalities are woven into everyday science learning. I find Bourdieu's concepts useful because they help me to think through how everyday practices lead to large-scale social patterns. Bourdieu (1984) and Bourdieu and Johnson (1993) argued that forms of political, educational and cultural participation are socially stratifying practices. In other words, they maintain social hierarchies by reproducing patterns of advantage and disadvantage through who can and cannot participate, a process Bourdieu called social reproduction (Bourdieu, 1990a; Bourdieu & Passeron, 1990).

Like Young, Bourdieu was interested in how social reproduction happens in subtle as well as obvious ways. His research explored how people could develop a taste for particular practices, and how these tastes might coalesce into social patterns across people's lifestyles. Bourdieu unpacked these ideas of taste and social reproduction to argue that social disadvantages (and advantages) are produced and maintained by the relationships between the structure of the situations people find themselves in (the field), the resources available to them (their capital) and people's behaviours and dispositions (their habitus) (Bourdieu, 1998). Much of the research on how participation in public activities reproduces social advantages and disadvantages focuses on the arts. I build here on Hesmondhalgh's (2006) reminder that Bourdieu's field of cultural production included law and science alongside the arts and extend these ideas to the various practices involved in everyday science learning.

Everyday science learning and the concept of field

How can we understand the importance of context in terms of exclusion and everyday science learning? Whether a specific field, or, as discussed in this book, a set of related fields, is accessible is clearly important, especially for those fields that are more dominant, or powerful, in our societies. As Young has argued, "some groups have exclusive or primary access to what Nancy Fraser (1987) calls the means of interpretation and communication in a society" (Fraser, 1987; Young, 1990, p. 58). Echoing Young and Fraser, Bourdieu argued that the more prestigious, or dominant, a given field is, the more inaccessible it is (Bourdieu & Johnson, 1993). Thus, what Miles and Gibson (2016) have called the orthodox view of cultural participation, positions activities along a spectrum from rarefied, high-brow and elite at one end to everyday, low-brow and popular at the other. As a result, participation within and across fields is understood to be a marker of privilege (Bennett et al., 2009; Khan, 2011; Thornton, 1996). Examining access to fields and the resources within them is therefore crucial for understanding how exclusion operates, as well as how inclusion might be achieved.

Bourdieu saw field as a "social universe having its own laws of functioning" (Bourdieu & Johnson, 1993, p. 14). These social universes are made up of people, each with resources, or forms of capital, which they use, accumulate or exchange according to how well they understand and can operate within that field as a result of their habitus (Bourdieu & Wacquant, 1992; Lareau & Horvat, 1999). Fields can be understood as the various social contexts you inhabit and, comprised as they are of the social positions of people and their practices, people typically engage with multiple fields in different ways over their lifetimes. Fields are, in this sense, social rather than purely physical spaces. To take an example from our research on the Enterprising Science project[2] for instance, we found visits to science museums worked differently for girls and boys (Archer et al., 2016; Dawson et al., 2019; Godec, 2018). Gender, 'race'/ethnicity, class and science were co-constructed in these spaces by both exhibits and signage, but also through the behaviours of teachers, museum facilitators and the students themselves. As a result, certain behaviours and knowledges that some boys brought to bear on their museum visit were rewarded, while the same was not true for the girls. Although they were in the same physical space, girls and boys had very different experiences – mediated by gender, 'race'/ethnicity and class – because the rules and values of that field created very different science learning experiences.

Everyday science learning happens in more than one field. Thus, as discussed in this book, participants' involvement in everyday science learning spanned home, school and beyond. Each of these fields has its own values and practices, or rules, and these serve to differentiate one field from the next. The social positions and possibilities within a field are therefore marked as much by those outside them as those inside. I take the view in this book that school, home, work, the mass media and the cultural industries (including museums, science centres, zoos but also popular science books) were all fields that influenced participants' everyday science learning.

Importantly, for Bourdieu, fields relate to other fields (Bourdieu & Johnson, 1993; Bourdieu & Wacquant, 1992). For example, the field of science museums is closely related to school science education, sharing aspects of the same political agendas, funding bodies and participants. You could expect therefore, that someone with capital in the field of school science education would be able to use that capital in a museum or science centre, and vice versa. To understand how field matters for equity, exclusion and everyday science learning therefore, it is important to think about the rules and values of a given field, relationships between fields and the possible positions within (and without) a field, as well as the resources, 'know-how' and dispositions required to navigate a field.

Capital and everyday science learning

Let's turn now to the resources – of forms of capital – that affect people's positions and possibilities within a field and how they might affect everyday science learning. Bourdieu argued that people navigate the settings they find themselves in (or fields) by using resources such as knowledge, skills, social connections, or money, which he called forms of capital, alongside their "feel for the game" or habitus (1998, p. 83). Crucially, for Bourdieu "a capital does not exist and function except in relation to a field" (Bourdieu & Wacquant, 1992, p. 101). This means that while capital can be exchanged and used between related fields as in the example of science museums and school science education above, this is not always the case because the value of capital is field and person dependent (as with the previous example of girls and boys at the museum). So, while currency is a form of economic capital and the economies of different countries are clearly related, a five-dollar note found by my two-year-old daughter on the pavement in the UK is nothing but paper in her hands. In other words, context (the combination of field and habitus) is crucial for understanding capital.

Bourdieu argued capital could be understood in many forms, but what I find most useful about the concept of capital as Bourdieu developed it is how it allows us to think about resources in a variety of tangible and less tangible ways. For instance, cultural capital – the form of capital I write about most in this book – can be understood as knowledge and familiarity with field-specific practices resulting in competency (Bourdieu & Darbel, 1991; Lamont & Lareau, 1988; Skeggs, 2004). It can be used, built or lost in specific contexts (fields) depending on how you have been socialised into that field through previous experiences (habitus).

I find the idea of capital helpful for thinking in detail about which kinds of resources, skills or literacies are needed or valued in different fields, in order to unpick how exclusion/inclusion works in that setting. In the context of everyday science learning for instance, studies have shown that what Pandora and Rader (2008, p. xx) called "the scientific imagination" is a valuable thinking practice. It can be considered 'science capital' and developed through taking part in everyday science learning practices (Archer, Dawson, DeWitt, Seakins, & Wong, 2015, p. 922; Kato-Nitta, 2013). In other words, scientific knowledge (whether about content, practices or applications) is a valuable form of cultural capital in our societies. Indeed, research suggests cultural capital (scientific literacy), social capital (contacts involved in science) and habitus (the extent to which people identify with science, feel comfortable taking part in science) all influence the development of science capital (Archer et al., 2015, 2012; DeWitt & Archer, 2017; DeWitt, Archer, & Mau, 2016; King & Nomikou, 2018).

Finally, it is also worth thinking about how identity resources can operate as forms of capital for everyday science learning in certain fields. Though ideas about identity are contested, thinking about social positions in terms of capital is useful, because as Holland, Skinner, Lachiotte Jr., and Cain (2001, p. 271) argue, "social position has to do with entitlement to social and material resources and so to the higher deference, respect, and legitimacy accorded to those genders, races, ethnic groups, castes, and sexualities privileged by society".

Bourdieu's concept of capital can be used to think about how who we are affects access to fields wherein forms of capital might be generated, exchanged or lost. As such, it can be used to think beyond the class analysis that characterised the majority of Bourdieu's work to consider the relationships between various social positions and capital. Indeed, as others have argued, capital, not least cultural capital, is inflected by 'race'/ethnicity, as well as gender (Moi, 1991; Puwar, 2009; Wallace, 2016).

Depending on the interplay of your social position, field, habitus and capital, you may be more or less able to accrue, use or exchange forms of capital, in ways that are marked by structural inequalities. Ghassan Hage (1998) has argued for instance, that whiteness works as a form of capital in certain cultural, political and educational fields, echoing studies of whiteness as a valued, respected resource in how publics are constructed (Puwar, 2004; Ware, 2015; Ware & Back, 2002). Similarly, Lisa Adkins and Bev Skeggs (2004) and Carrie Paechter (2007) note that gender can also be considered a form of capital depending on the field. For instance, being male carries with it a certain amount of symbolic capital in science education and careers, that disadvantages women with similar aptitudes, skills and qualifications, in ways that are also marked by 'race'/ethnicity and class (Dawson et al., 2019; Ong, Wright, Espinosa, & Orfield, 2011). This brings us back to thinking about structural inequalities and Young's (1990) and Fraser's (1987) previous point, that some groups enjoy more access to certain fields and resources – or forms of capital – than others, and indeed, by virtue of who they are, possess more valued forms of symbolic capital to begin with. Thus, to understand issues of equity and exclusion, we need to understand the relationships between social positions and those forms of capital that people might accrue and exchange through their participation in everyday science learning practices.

Everyday science learning as habitus

Habitus is the next piece of the puzzle for understanding Bourdieu's theories of social reproduction in relation to everyday science learning, equity and exclusion. Bourdieu described habitus as the result of people's experiences, the "conditions of existence which, in imposing different definitions of the impossible, the possible, and the probable, cause one group to experience as natural or reasonable practices or aspirations which another group finds unthinkable or scandalous, and vice versa" (1977, p. 78). In this book I use the concept of habitus to try to understand how people recognise their worlds, such that without thinking they know what to pay attention to, what to overlook, or what to do in particular situations.

Bourdieu argued that through habitus, experiences of (dis)enfranchisement are mirrored in "different sets of dispositions with regard the social games that are held to be crucial to society" (Bourdieu & Wacquant, 1992, p. 172). Thus people become disposed against practices they feel are of little use or relevance to them, for example, choosing not to study science at university because they do not see it as useful for the kinds of jobs they expect to do in the future. This "system of dispositions" is, however, dynamic (Bourdieu & Passeron, 1990, p. 67). Habitus affects new experiences as much as it is influenced by earlier experiences and, while resilient, it is not fixed. Thus, while a person's habitus structures their behaviours, assumptions and lifestyle, it could shift over time as the result of transformative experiences and field-wide structural changes.

Bourdieu argued that patterns of habitus can be identified within social groups whose experiences may be shared. People exposed to similar settings, attitudes and practices share a similar habitus, because, as Bourdieu and Wacquant argue, these experiences are "already predefined by broader racial, gender, and class relations" (1992, p. 144). Thus, Bourdieu argued that people in similar social positions share similar experiences, which in turn produce a similar habitus. Bourdieu and Passeron called this "class habitus" (1990, p. 204), an idea that Umut Erel (2010) argues can be extended to racialised groups with sufficient shared experiences. For Bourdieu and Darbel (1991), the conceptual art museum was the preserve of the upper classes and would only ever have a limited appeal to working-class people who may perceive such institutions as irrelevant for them. If this is the case with class habitus and art galleries, to what extent is this also the case for people from racialised groups and their attitudes to science or their avoidance of science centres?

Social justice, Bourdieu and exclusion

In this last section I turn to how we can understand the damage done if you are not able to build, use or exchange capital or even access a particular everyday science learning field to begin with. Social justice theories, though much debated and discussed, can for our purposes, be grouped into three types: those that focus on access and the equal distribution of resources (redistributive social justice), those that focus on equity, on recognising and respecting people's differences (relational social justice) and those that combine both approaches (Fraser & Honneth, 2003; Rawls, 1971; Young, 1988, 1990, 2000).

Questions of access are fundamental social justice concerns and often not so hard to unpack. Not being able to access or use a resource is a clear form of exclusion. In contrast, what it means to be excluded in subtle ways though not being respected, represented or even recognised can be harder to pin down. Two forms of oppression discussed by Young (1988, 1990) – cultural imperialism and powerlessness – are particularly helpful concepts for thinking through the more subtle aspects of exclusion from everyday science learning. Bourdieu's (1990b) idea of symbolic violence is useful for understanding how exclusion might masquerade as a choice for some people in some contexts.

Cultural imperialism, for Young (1990), is experienced when socially dominant perspectives and practices suppress or invalidate the views and experiences of minoritised groups, rendering them invisible. Building on the work of Edward Said, Young argued that "cultural imperialism involves the universalisation of a dominant group's experience and culture, and its establishment as the norm" (Said, 1993; Young, 1990, p. 58). Cultural imperialism works to benefit dominant groups. This can be seen in how McIntosh (1989) formulated her view of white privilege. She argued that white privilege meant seeing people like you represented in popular media and stories about history in a positive way, and that these plentiful and positive representations were utterly normal. From the perspective of minoritised groups, however, cultural imperialism can be seen in everyday science learning when cultural artefacts and histories are displayed in an appropriative way in ethnographic exhibits without their co-operation and in ways that mark their knowledge, practices and selves as Other (Lavine & Karp, 1991; Onciul, 2015).

Powerlessness, as developed by Young (1990), combines issues of 'race'/ethnicity, gender and class to describe the experience of being disrespected and having little or no autonomy over your choices, for instance, in terms of employment or political voice because of your marginalised social status. An example of this in everyday science learning is when people are not listened to in a socio-scientific consultation exercise or when their opinions are not even sought. Powerlessness can also be seen when, even if invited to participate, the contexts of people's lives are such that they could not take up that invitation.

The final idea I turn to in this section is drawn from Bourdieu's (1990b, 1991) work – symbolic violence – and is helpful for thinking about how a practice like non-participation in certain everyday science learning fields could seem like a choice, as much as the result of exclusive practices. For Bourdieu, symbolic violence lay in the misrecognition of overt exclusion, domination or inherited advantage. He described it as "gentle, invisible violence, unrecognized as such, chosen as much as undergone" (Bourdieu, 1990b, p. 127). For example, a school system where the knowledge that counts is that of the dominant groups can be understood to disadvantage students from non-dominant groups by rendering their knowledges and practices as Other. Those students may decide not to pursue higher education, taking themselves out of the system that excludes them (Bourdieu & Passeron, 1990; Ogbu, 1992; Yosso, 2005; Young, 2000). Symbolic violence is the misrecognition of power and agency, such that the disenfranchised – the working class for Bourdieu and Passeron – make a virtue of necessity by interpreting inaccessible opportunities as choices not to participate. That such practices obscure the reproduction of inequality is at the heart of their power (Bourdieu & Passeron, 1990).

These three concepts – cultural imperialism, powerlessness and symbolic violence – are useful because they give us a handle on how exclusion from everyday science learning operates. As Puwar (2004) has argued, it is crucial that we learn more about how exclusion is understood and experienced by those at the sharp end of such practices if we are to stand a chance of working towards inclusion and equity. Thus, across the remaining chapters of this book, I explore participants' experiences of and attitudes towards everyday science learning.

Summary

In this chapter I argued that the ways in which exclusion and inclusion are currently (mis)understood in everyday science learning do not and cannot support the development of more meaningfully inclusive practices. I discussed how framing exclusion in terms of deficits while framing inclusion in terms of an assimilationist crusade works to shore up the status quo in favour of dominant groups. In the second part of the chapter I outlined the theoretical approach I take in this book as an alternative way to think about exclusion and, by extension, inclusion. I argued that we must move beyond deficit approaches to understanding exclusion and beyond crusade approaches to understanding inclusion if we are to disrupt and transform everyday science learning and redevelop inclusive, equitable practices.

Notes

1 In this book I focus on the mediating practices between science and publics in everyday science learning practices, but there is a significant, larger question about their content that needs also to be taken into account. In other words, there is a 'science' question to be taken seriously from an equity and inclusion perspective in terms of what exactly it is we are seeking to include people in. It is because I have been too long involved in science and technology studies that I can no longer accept a simple view that we need to solve inclusion 'into' science. Rather, as I discuss in more detail in Chapter Eight, the disruptive, transformative project I outline for equity and inclusion in everyday science learning necessarily includes disrupting and transforming science too.

2 Enterprising Science was a research and development project partnership between University College London, King's College London and the Science Museum that sought to explore and develop the concept of science capital, and was led by Louise Archer.

References

Adkins, L., & Skeggs, B. (2004). *Feminism after Bourdieu*. Oxford and Malden, MA: Blackwell Publishing.

Ahmed, S. (2012). *On being included: Racism and diversity in institutional life*. Durham and London: Duke University Press.

Archer, L., Dawson, E., DeWitt, J., Seakins, A., & Wong, B. (2015). "Science capital": A conceptual, methodological, and empirical argument for extending bourdieusian notions of capital beyond the arts. *Journal of Research in Science Teaching*, 52, 922–948. doi.org/10.1002/tea.21227

Archer, L., Dawson, E., Seakins, A., DeWitt, J., Godec, S., & Whitby, C. (2016). "I'm being a man here": Urban boys' performances of masculinity and engagement with science during a science museum visit. *Journal of the Learning Sciences*, 25(3), 438–485. doi:10.1080/10508406.2016.1187147

Archer, L., DeWitt, J., Osborne, J., Dillon, J., Willis, B., & Wong, B. (2012). Science aspirations, capital, and family habitus: How families shape children's engagement and identification with science. *American Educational Research Journal*, 49(5), 881–908. doi:10.3102/0002831211433290

Association of Science and Discovery Centres. (2014). *UK science and discovery centres: Effectively engaging under-represented groups*. Bristol: Association of Science & Discovery Centres.

Association of Science and Technology Centres. (1987). *Natural partners: How science centres and community groups can team up to increase scientific literacy*. Washington, DC: Association of Science and Technology Centres.

Bassel, L., & Emejulu, A. (2017). *Minority women and austerity: Survival and resistence in France and Britain*. Bristol: Polity Press.

Belfiore, E. (2009). On bullshit in cultural policy practice and research: Notes from the British case. *International Journal of Cultural Policy, 15*(3), 343–359. doi:10.1080/10286630902806080

Bell, A. R., Davies, S. R., & Mellor, F. (2008). *Science & its publics*. Newcastle: Cambridge Scholars Publishing.

Bell, P., Lewenstein, B., Shouse, A. W., & Feder, M. A. (2009). *Learning science in informal environments: People, places, and pursuits*. Washington, DC: The National Academies Press.

Bennett, T., Savage, M., Silva, E., Warde, A., Gayo-Cal, M., & Wright, D. (2009). *Culture, class, distinction*. Abingdon and New York: Routledge.

Bhopal, K. (2018). *White privilege: The myth of a post-racial society*. Bristol: Polity Press.

Bourdieu, P. (1977). *Outline of a theory of practice* (R. Nice, Trans.). Cambridge: Cambridge University Press.

Bourdieu, P. (1984). *Distinction: A social critique of the judgement of taste* (R. Nice, Trans.). London: Routledge.

Bourdieu, P. (1990a). *In other words: Essays towards a reflexive sociology*. Stanford: Stanford University Press.

Bourdieu, P. (1990b). *The logic of practice* (R. Nice, Trans.). Stanford: Stanford University Press.

Bourdieu, P. (1991). *Language and symbolic power* (G. Raymond & M. Adamson, Trans.). Cambridge and Malden, MA: Polity Press.

Bourdieu, P. (1998). *Practical reason*. Cambridge: Polity Press.

Bourdieu, P., & Darbel, A. (1991). *The love of art: European art museums and their public*. Oxford: Polity Press.

Bourdieu, P., & Johnson, R. (1993). *The field of cultural production: Essays on art and literature*. Cambridge: Polity Press.

Bourdieu, P., & Passeron, J.-C. (1990). *Reproduction in education, society and culture* (R. Nice, Trans., 2nd ed.). London, Newbury Park, CA and New Delhi: Sage.

Bourdieu, P., & Wacquant, L. (1992). *An invitation to reflexive sociology*. Chicago: University of Chicago Press.

Cassidy, A., Lock, S. J., & Voss, G. (2016). Sexual nature? (Re)presenting sexuality and science in the museum. *Science as Culture, 25*(2), 214–238. doi:10.1080/09505431.2015.1120284

Crenshaw, K. (1991). Mapping the margins: Intersectionality, identity politics, and violence against women of color. *Stanford Law Review, 43*, 1241–1299.

Dahl, R. (1964). *Charlie and the chocolate factory*. London: Penguin Books.

Davies, S. R., & Horst, M. (2016). *Science communication: Culture, identity and citizenship*. London: Palgrave Macmillan.

Dawson, E. (2018). Reimagining publics and (non)participation: Exploring exclusion from science communication through the experiences of low-income, minority ethnic groups. *Public Understanding of Science, 27*(7) 772–786. doi:10.1177/0963662517750072

Dawson, E., Archer, L., Seakins, A., DeWitt, J., Godec, S., King, H., Mau, A. & Nomikou, E. (2019). Selfies at a science museum: Exploring girls' identity performances in a science learning setting. *Gender and Education*.

Department for Culture Media and Sport. (2016). *Taking part: Longitudinal report 2016*. London: Department for Culture, Media & Sport.

Dewey, J. (1927). *The public and its problems: An essay in political inquiry*. New York: Holt.

DeWitt, J., & Archer, L. (2017). Participation in informal science learning experiences: The rich get richer? *International Journal of Science Education, Part B*, 7(4), 356–373. doi:10.1080/21548455.2017.1360531

DeWitt, J., Archer, L., & Mau, A. (2016). Dimensions of science capital: Exploring its potential for understanding students' science participation. *International Journal of Science Education*, 38(16), 2431–2449. doi:10.1080/09500693.2016.1248520

Ebrey, J. (2016). The mundane and insignificant, the ordinary and the extraordinary: Understanding everyday participation and theories of everyday life. *Cultural Trends*, 25(3), 158–168. doi.org/10.1080/09548963.2016.1204044

Erel, U. (2010). Migrating cultural capital: Bourdieu in migration studies. *Sociology*, 44(4), 642–660.

Falk, J. (2009). *Identity and the museum visitor experience*. Walnut Creek: Left Coast Press.

Falk, J., Dierking, L. D., Osborne, J., Wenger, M., Dawson, E., & Wong, B. (2015). Analyzing science education in the United Kingdom: Taking a system-wide approach. *Science Education*, 99(1), 145–173.

Falk, J., Osborne, J., Dierking, L. D., Dawson, E., Wenger, M., & Wong, B. (2012). *Analyzing the UK Science Education Community: The contribution of informal providers*. London: Wellcome Trust.

Falk, J., & Storksdieck, M. (2010). Science learning in a leisure setting. *Journal of Research in Science Teaching*, 47(2), 194–212. doi:10.1002/tea.20319

Feinstein, N. W., & Meshoulam, D. (2014). Science for what public? Addressing equity in American science museums and science centers. *Journal of Research in Science Teaching*, 51(3), 368–394. doi.org/10.1002/tea.21130

Fenichel, M., & Schweingruber, H. A. (2010). *Surrounded by science: Learning science in informal environments*. Washington, DC: The National Academies Press.

Fourcade, M., & Healy, K. (2016). Seeing like a market. *Socio-Economic Review*, 15(1), 9–29. doi:10.1093/ser/mww033

Fraser, N. (1987). Social movements vs. disciplinary bureaucracies: The discourse of social needs. *CHS Occasional Paper* (Vol. No. 8). Minneapolis & St Paul: Centre for Humanistic Studies, University of Minnesota.

Fraser, N. (2003). Social justice in the age of identity politics: Redistribution, recognition, and participation. In N. Fraser & A. Honneth (Eds.), *Redistribution or recognition? A political-philosophical exchange* (pp. 7–109). London and New York: Verso.

Fraser, N., & Honneth, A. (2003). *Redistribution or recognition? A political-philosophical exchange*. London and New York: Verso.

Gillborn, D. (2010). The colour of numbers: Surveys, statistics and deficit-thinking about race and class. *Journal of Education Policy*, 25(2), 253–276.

Gillborn, D., Warmington, P., & Demack, S. (2018). QuantCrit: Education, policy, "Big Data" and principles for a critical race theory of statistics. *Race Ethnicity and Education*, 21(2), 158–179. doi:10.1080/13613324.2017.1377417

Gilroy, P. (2002). *There ain't no Black in the Union Jack* (2nd ed.). Abingdon: Routledge.

Godec, S. (2018). Sciencey girls: Discourses supporting working-class girls to identify with science. *Education Sciences, 8*(1), 1–17.

Golding, V. (2009). *Learning at the museum frontiers: Identity, race and power.* Farnham and Burlington: Ashgate Pub. Co.

Gunaratnam, Y. (2015). *Intersectional pain: What I've learned from hospices and feminism of colour.* Retrieved from https://www.opendemocracy.net/transformation/yasmin-gunaratnam/intersectional-pain-what-i've-learned-from-hospices-and-feminism-of-colour.

Gurian, E. H. (2006). *Civilizing the museum: The collected writings of Elaine Heumann Gurian.* Abingdon and New York: Routledge.

Hage, G. (1998). *White nation: Fantasies of White supremacy in a multicultural society.* New York: Routledge.

Harding, S. (2006). *Science and social inequality: Feminist and postcolonial issues.* Urbana and Chicago: University of Illinois Press.

Hesmondhalgh, D. (2006). Bourdieu, the media and cultural production. *Media, Culture & Society, 28*(2), 211–231.

Hill Collins, P., & Bilge, S. (2016). *Intersectionality.* Cambridge: Polity Press.

Holland, D., Skinner, D., Lachiotte Jr., W., & Cain, C. (2001). *Identity and agency in cultural worlds.* Cambridge, MA and London: Harvard University Press.

Ipsos MORI. (2003). *The impact of free entry to museums.* London: Ipsos MORI.

Ipsos MORI. (2014). *Public attitudes to science 2014.* London: Department for Business, Innovation and Skills.

Ipsos MORI. (2016). *Wellcome Trust monitor report wave 3: Tracking public views on science and biomedical research.* London: Wellcome Trust.

Kato-Nitta, N. (2013). The influence of cultural capital on consumption of scientific culture: A survey of visitors to an open house event at a public scientific research institution. *Public Understanding of Science, 22*(3), 321–334.

Khan, S. R. (2011). *Privilege: The making of an adolecent elite at St. Paul's School.* Princeton, NJ and Oxford: Princeton University Press.

King, H., & Nomikou, E. (2018). Fostering critical teacher agency: The impact of a science capital pedagogical approach. *Pedagogy, Culture & Society, 26*(1), 87–103. doi:10.1080/14681366.2017.1353539

Lamont, M., & Lareau, A. (1988). Cultural capital: Allusions, gaps and glissandos in recent theoretical developments. *Sociological Theory, 6*(2), 153–168.

Lareau, A., & Horvat, E. M. (1999). Moments of social inclusion and exclusion race, class, and cultural capital in family-school relationships. *Sociology of Education, 72*(1), 37–53. doi:10.2307/2673185

Lavine, S. D., & Karp, I. (1991). Introduction: Museums and multiculturalism. In I. Karp & S. D. Lavine (Eds.), *Exhibiting cultures: The poetics and politics of museum display* (pp. 1–10). Washington, DC: Smithsonian Institution.

Lee, O. (1999). Equity implications based on the conceptions of science achievement in major reform documents. *Review of Educational Research, 69*(1), 83–115.

Lee, O., & Buxton, C. A. (2010). *Diversity and equity in science education: Research, policy, and practice.* New York: Teachers College Press.

Leong, N. (2013). Racial capitalism. *Harvard Law Review, 126*(8), 2153–2226.

Levitas, R. (1998). *The inclusive society?* Basingstoke and New York: Palgrave Macmillan.

Levitas, R. (2004). Let's hear it for Humpty: Social exclusion, the third way and cultural capital. *Cultural Trends, 13*(2), 41–56.

Lock, S. J. (2011). Deficits and dialogues: Science communication and the public understanding of science in the UK. In D. J. Bennett & R. C. Jennings (Eds.), *Successful science communication: Telling it like it is* (pp. 17–30). Cambridge: Cambridge University Press.

Lynch, B. (2011). *Whose cake is it anyway? A collaborative investigation into engagement and participation in 12 museums and galleries in the UK.* London: Paul Hamlyn Foundation.

Marris, C. (2014). The construction of imaginaries of the public as a threat to synthetic biology. *Science as Culture, 24*(1), 83–98.

McGuigan, J. (1996). *Culture and the public sphere.* London and New York: Routledge.

McIntosh, P. (1989, July/August). White privilege: Unpacking the invisible knapsack. *Peace and Freedom,* pp. 10–12.

Medin, D. L., & Bang, M. (2014). *Who's asking? Native science, western science, and science education.* Cambridge, MA and London: MIT Press.

Michael, M. (2012). "What are we busy doing?": Engaging the idiot. *Science Technology & Human Values, 37*(5), 528–554.

Miles, A., & Gibson, L. (2016). Everyday participation and cultural value. *Cultural Trends, 25*(3), 151–157.

Moi, T. (1991). Appropriating Bourdieu: Feminist theory and Pierre Bourdieu's sociology of culture. *New Literary History, 22*(4), 1017–1049. doi:10.2307/469077

Newman, A., McLean, F., & Urquhart, G. (2005). Museums and the active citizen: Tackling the problems of social exclusion. *Citizenship Studies, 9*(1), 41–57. doi:10.1080/1362102042000325351

Ogbu, J. U. (1992). Understanding cultural diversity and learning. *Educational Researcher, 21*(8), 5–24.

Onciul, B. (2015). *Museums, heritage and indigenous voice.* London and New York: Routledge.

Ong, M., Wright, C., Espinosa, L. L., & Orfield, G. (2011). Inside the double bind: A synthesis of empirical research. *Harvard Educational Review, 81*(2), 172–390.

Packer, J. (2008). Beyond learning: Exploring visitors' perceptions of the value and benefits of museum experiences. *Curator: The Museum Journal, 51*(1), 33–54. doi. org/10.1111/j.2151-6952.2008.tb00293.x

Paechter, C. (2007). *Being boys, being girls: Learning masculinities and femininities.* Maidenhead: Open University Press.

Pandora, K., & Rader, A. K. (2008). Science in the everyday world. *Isis, 99*(2), 350–364.

Phipps, A. (2008). *Women in science, engineering, and technology: Three decades of UK initiatives.* Sterling, VA: Trentham Books.

Pollock, A., & Subramaniam, B. (2016). Resisting power, retooling justice: Promises of feminist postcolonial technosciences. *Science, Technology & Human Values, 41*(6), 951–966. doi:10.1177/0162243916657879

Puwar, N. (2004). *Space invaders: Race, gender and bodies out of place.* Oxford and New York: Berg.

Puwar, N. (2009). Sensing a post-colonial Bourdieu: An introduction. *The Sociological Review, 57*(3), 371–384. doi:10.1111/j.1467-954X.2009.01856.x

Rawls, J. (1971). *A theory of justice.* Cambridge, MA: Harvard University Press.

Rennie, L. J., & Williams, G. F. (2006). Adults' learning about science in free-choice settings. *International Journal of Science Education, 28*(8), 871–893.

Rommetveit, K., & Wynne, B. (2017). Technoscience, imagined publics and public imaginations. *Public Understanding of Science, 26*(2), 133–147.

Said, E. W. (1993). *Culture and imperialism.* London: Vintage.

Sandell, R. (1998). Museums as agents of social inclusion. *Museum Management and Curatorship, 17*(4), 401–418.

Sandell, R. (2007). *Museums, prejudice and the reframing of difference.* London and New York: Routledge.

Savage, M. (2010). *Identities and social change in Britain since 1940: The politics of method.* Oxford and New York: Oxford University Press.

Schiebinger, L. (2007). Getting more women into science: Knowledge issues. *Harvard Journal of Law and Gender, 30*, 365–378.

Scott, S. (2009). *Making sense of everyday life.* Cambridge: Polity Press.

Skeggs, B. (2004). *Class, self, culture.* London and New York: Routledge.

Solomon, J. (2012). *Science of the people: Understanding and using science in everyday contexts.* New York and Abingdon: Routledge.

Stanley, W. B., & Brickhouse, N. W. (2001). Teaching sciences: The multicultural question revisited. *Science Education, 85*(1), 35–49.

Sturgis, P., & Allum, N. (2004). Science in society: Re-evaluating the deficit model of public attitudes. *Public Understanding of Science, 13*(1), 55–74.

Taylor, M. (2016). Taking part: The next five years (2016) by the Department for Culture, Media and Sport. *Cultural Trends, 25*(4), 291–294.

Taylor, M., & O'Brien, D. (2017). "Culture is a meritocracy": Why creative workers' attitudes may reinforce social inequality. *Sociological Research Online, 22*(4), 27–47. doi:10.1177/1360780417726732

Thornton, S. (1996). *Club cultures: Music, media, and subcultural capital.* Hanover, NH: University Press of New England.

Tlili, A. (2008). Behind the policy mantra of the inclusive museum: Receptions of social exclusion and inclusion in museums and science centres. *Cultural Sociology, 2*(1), 123–147.

Tlili, A., & Dawson, E. (2010). Mediating science and society in the EU and UK: From information-transmission to deliberative democracy? *Minerva, 48*(4), 429–461.

Tunnicliffe, S. D. (2008). Conversations of family and primary school groups at robotic dinosaurs in a museum? What do they talk about? *Journal of Elementary Science Education, 20*(3), 17–33.

Turney, J. (1998). To know science is to love it? Observations from public understanding of science research. London: The Royal Society.

Wallace, D. (2016). Reading "race" in Bourdieu? Examining Black cultural capital among Black Caribbean Youth in South London. *Sociology, 51*(5), 907–923. 0038 038516643478. doi:10.1177/0038038516643478

Ware, V. (2015). *Beyond the Pale* (2nd ed.). London: Verso.

Ware, V., & Back, L. (2002). *Out of whiteness: Color, politics, and culture.* Chicago and London: University of Chicago Press.

Wynne, B. (1992). Misunderstood misunderstanding: Social identities and public uptake of science. *Public Understanding of Science, 1*(3), 281–304. doi:10.1088/0963-6625/1/3/004

Yosso, T. J. (2005). Whose culture has capital? A critical race theory discussion of community cultural wealth. *Race Ethnicity and Education, 8*(1), 69–91.

Young, I. M. (1988). Five faces of oppression. *The Philosophical Forum, 19*(4), 270–290.

Young, I. M. (1990). *Justice and the politics of difference.* Princeton, NJ: Princeton University Press.

Young, I. M. (2000). *Inclusion and democracy.* Oxford and New York: Oxford University Press.

Chapter 3

Mapping participation

"Hawking hails London, capital of science", read the front-page headline of the *Evening Standard* newspaper above an article about a science-themed party for the "1000 power list" of influential people (Randhawa & Spanier, 2014, p. 1). Places like London include astonishing opportunities not only for scientific research, but for everyday science learning. For instance, London's largest, new scientific research centre, The Francis Crick Institute, moved into lavish, newly built laboratories next to the British Library in 2016. It represents a collaboration between the UK's Medical Research Council, Cancer Research UK, University College London, Imperial College London and King's College London and, as such, is one of the largest biomedical research sites in the country. London is also home to 12 of the 14 nationalised museums in England, including what are arguably the two most famous science museums in the UK, the Natural History Museum and the Science Museum, as well as five others with science-related collections/content.[1] In many ways London really is a capital of science. But what does science mean for people in London with little or no access to the various fields involved in everyday science learning, let alone parties with science celebrities?

A driving assumption across the various fields involved in research on science and society relationships is that more science communication, engagement and learning is a good thing. While participation in everyday science learning can have both societal and individual benefits, these benefits are contested. Thus, at the societal level, everyday science learning can be seen as part of deliberative, participatory democracy that improves public knowledge of and involvement in science, thereby improving policy decisions, while also criticised as simply a capitalist feature of selling science (Stilgoe, Lock, & Wilsdon, 2014; Thorpe & Gregory, 2010). Non-participation is problematic for societies in a normative sense therefore, since it may impair political and market processes. At the personal level, participation in everyday science learning practices are thought to benefit people by sharing valuable knowledge, by opening up science policy practices or as a key element of our culture (Davies & Horst, 2016; Delgado, Lein Kjølberg, & Wickson, 2011). Questions remain however, at both scales, about whose values, knowledge and culture such practices reproduce, as well as how accessible they are.

To paraphrase Mariana Ortega (2006), questions remain about whose experiences, practices and knowledges we attend to in research about everyday science learning. This partial focus affects knowledge making in that research and related policies or practices. In this chapter and those that follow I use empirical data to try to provide an alternative perspective on everyday science learning, the perspective of those underserved by that system. I begin by mapping broad patterns about who participates in science communication and informal science learning drawing on the available literature about these two overlapping but not entirely similar areas of practice. I also argue that we need to do more than simply count who does what if we want to understand equity, exclusion and everyday science learning. I then map patterns of cultural consumption in relation to everyday science learning for participants involved in the research for this book. While mapping cultural consumption is frequently done in cultural studies to explore participation in arts and cultural practices, it is rarely used to look at participation in science-related cultural practices (Archer, Dawson, DeWitt, Seakins, & Wong, 2015; Bennett et al., 2009; Prieur & Savage, 2013). My driving question here is whether the fields associated with everyday science learning are accessible.

As a point of comparison, in this chapter I also map participants' patterns of cultural consumption more broadly. As I learnt through my time with them, many participants were seasoned political activists on behalf of their communities and almost all were steeped in what Sandra Trienekens (2002, p. 283) has called "community-based" cultural capital. That is, they were frequent participants in community-based cultural activities, both as producers and audiences/users. Building on the arguments made in Chapter Two, I use this data to argue that we should be wary of accidentally slipping into a deficit approach to understanding participation in everyday science learning since participants clearly led rich cultural lives.

Through mapping cultural consumption I argue that access issues beset the fields involved in everyday science learning. These issues become particularly visible in comparison to how much participants were involved in other cultural, educational and political activities. I discuss how the pattern revealed by mapping participants' everyday science learning practices shows these are not only the result of individual choices and preferences, but that everyday science learning can be seen as comprising restricted and thereby exclusive fields of cultural production (Bourdieu & Johnson, 1993). I argue in this chapter that it is inadequate to consider participants in terms of participatory deficit when it comes to cultural, educational or political activity. Instead, we need to take the idea that everyday science learning involves restricted and exclusive fields of cultural production more seriously. In other words, I found that through no fault of the participants, everyday science learning was inaccessible.

Counting bodies, bodies that count

Who participates in everyday science learning as an audience, user or visitor? And what do we already know about how patterns of participation and access play out across populations? In this section I use descriptive survey data to look at what we already know about who does and, importantly, who does not participate in everyday science learning. I do so however with a series of provisos. First, I wholeheartedly agree with Audre Lorde (1984), that calls for more data can be used as a strategy to delay or resist change. I have heard this as excuse too many times throughout this project. We already have enough data about patterns of participation in everyday science learning to know that things are far from fair, as this section demonstrates. Second, as many scholars have argued, counting bodies is rarely a sufficient requirement for or description of equity and inclusion (Leong, 2013; Puwar, 2004; Saha, 2018). I do not mean by this that we should stop counting who does what, but rather that we should be wary of claims made about inclusive practice on the basis of numbers alone. Numbers provide an illustrative backdrop against which we must ask more useful questions.

Everyday science learning practices can, in many ways, be considered as growing and popular pursuits. Millions of people visit science centres, museums, zoos, botanic gardens, science festivals and more around the world (National Science Foundation, 2012; OECD, 2012). Data compiled in 2012, for instance, show that between a fifth (22% in Brazil) and half (52% in China and the United States) of people in China, Japan, South Korea, India, Malaysia, the United States, the European Union, and Brazil visited zoos, aquaria and science museums (National Science Foundation, 2012). Sally Duensing's (2006) qualitative, long-term study of science centre practices in different countries suggests that while local practices may differ, the notion of a science centre has spread internationally. Similarly, although international research on science festivals suggests that although science festivals remain concentrated in Europe, there is significant international growth in these events, with more than 5.6 million people worldwide taking part (Bultitude, McDonald, & Custead, 2011; Wiehe, 2014).

In the UK, where the research for this book took place, a 2014 report on public views of science found that of the people surveyed "two-thirds (67%) have been to at least one of the science-related leisure or cultural activities asked about in the survey in the previous year" (Ipsos MORI, 2014, p. 109). The activities asked about were science talks and activities outside of school or university classes, visits to science centres, science museums, zoos, aquaria, nature reserves, planetaria and science festivals. Of these activities, nature reserves, zoos and aquaria, museums and science centres were, in that order, the most popular (Ipsos MORI, 2014). While the inclusion of nature reserves as a category in the 2014 survey increased the total proportion of participating adults, these data still support the results of an earlier version of the same survey that found over half the surveyed public were involved in the science activities the survey asked about (Ipsos MORI, 2011). Although these kinds of everyday science learning activities were popular, television and school remained the main sources of scientific information for those surveyed (Ipsos MORI, 2014, 2016). Thus, these data suggest people engage with science across different fields, in different ways.

Despite these activities' apparent popularity, data about who participates suggests the appeal of everyday science learning is not as broad as might be hoped.

Relatively little is known about why and how patterns of non-participation in everyday science learning exist and persist, a concern raised by others in the UK and echoed across Europe and the US (Atkinson, Siddall, & Mason, 2014; Bell, Lewenstein, Shouse, & Feder, 2009; Massarani & Merzagora, 2014; The Association for Science and Discovery Centres, 2010; Wellcome Trust, 2008). Patterns of non-participation can, however, be inferred by exploring who is missing from data on participation in various kinds of everyday science learning. We should remember though that the available data reflects decisions about whose practices are important enough to research and, as a result, tend to reflect orthodox approaches to cultural participation and deficit approaches to racialised and working-class groups (Gillborn, 2010; Gillborn, Warmington, & Demack, 2018; Miles & Gibson, 2016; Taylor, 2016). As a result, we know more about who visits science centres than we do about who takes part in community-led science activities or watches science-related television at home, and that knowledge is marked by structural inequalities.

Staying with the UK as an example, most science centre visitors are middle-class people, particularly schoolchildren, from white British ethnic backgrounds (Ecsite-UK, 2008; Ipsos MORI, 2011, 2014; Wellcome Trust, 2008). Research on museum visitors in the UK highlights similar patterns. In terms of the bodies that count, what this means is that in the British context, everyday science learning participants are drawn from White ethnic backgrounds, middle and upper classes, live in cities and use everyday science learning opportunities with their families or schools (Department for Culture Media and Sport, 2011, 2016; Ipsos MORI, 2001, 2006, 2014). This pattern is echoed in international data about who participates in everyday science learning (National Science Foundation, 2012; OECD, 2012).

Clearly a broad range of practices, places and professionals co-exist within everyday science learning. However, research suggests the same kinds of people access and use these different resources time and time again (Dawson, 2014; DeWitt & Archer, 2017; Falk et al., 2015; Feinstein, 2017). Thus, while the visitor or audience profiles of individual science festivals, zoos or science museums may vary, taking a step back to look across the system more broadly suggests these institutions and the resources they hold are marked by structural inequalities. Instead of providing valuable science-related opportunities for everyone, everyday science learning practices seem, instead, to be predominantly used by the more privileged groups of our societies.

Mapping participation from the available descriptive data suggests therefore that access to the fields of everyday science learning about which we have data are socially stratified. Those in the more advantaged social positions are over represented amongst visitor, audience or user numbers. To look at this picture in reverse, these data suggest that people from lower socio-economic groups, from racialised groups, older adults, those living in rural areas and away from their families are unlikely to use everyday science learning resources. Exclusion from everyday science learning therefore appears hierarchical (because it is patterned by social disadvantage) and intersectional (through the combined effects of multiple structural inequalities).

Mapping patterns of participation in everyday science learning and, by inference, patterns of non-participation highlights how what might be everyday for some is far from everyday for others (Scott, 2009). At the same time however, I do not want to suggest that these numbers provide enough insight into how exclusion from everyday science learning is experienced, nor insight into what inclusion would require. As Puwar (2004, p. 32) has argued "our analysis needs to go beyond number-crunching exercises which count (monitor) the quantities of different bodies in the stratified structures of institutions". Instead, we might turn our attention more usefully to experiences of everyday science learning and exclusion from the perspective of those who are excluded, as I do in the rest of this chapter.

Cultural consumption and everyday science learning

Exploring patterns of cultural consumption in everyday science learning is useful because it highlights access issues within and between fields. In this section I briefly discuss how cultural consumption can be understood in terms of everyday science learning and what that might mean in the context of exclusion and non-participation.

Accessibility and the redistribution of access to fields and the resources within them is crucial for thinking about equity, exclusion and everyday science learning. For Bourdieu and Johnson (1993), restricted access to production and participation in a given field is a sign of the prestige and value associated with that field. In other words, exclusive, inaccessible fields are more dominant in our societies and participation in them confers more advantage. Andrew Miles and Lisanne Gibson refer to this perspective as the orthodox or " 'official' framework of cultural participation and value in the UK" (2016, p. 151) since it reflects classed, gendered and racialised assumptions about how certain fields of practice are valued while others are not. Although everyday science learning includes practices that can be understood as educational or political as well as cultural, exploring patterns of participation in terms of cultural consumption from the perspective of those who do not participate helps us to see where access issues arise.

Thinking about who participates in which activities as well as *how* they participate is crucial for understanding how participation is involved in the social reproduction of privilege and disadvantage. Scholars have framed cultural practices along a spectrum of social privilege, as, for instance high-brow or low-brow, elite or popular, traditional or everyday. Of course distinctions between elite and popular culture blur, and as Peter Broks (2006, p. 1) reminds us, "the culture we live in is saturated with science". But how might exclusion and equity issues play out in relation to science across different cultural forms? Typically, high-brow, elite or traditional practices are more restricted, exclusive, powerful and use dominant forms of capital, particularly when it comes to cultural capital (Lareau & Weininger, 2003; McGuigan, 1996). In thinking about everyday science learning, this would mean that visiting a science centre or taking part in a government consultation about a socio-scientific issue could be seen as more prestigious, and could confer more advantages, than watching a YouTube film about science at home.

Scholars have also argued that social advantage is reproduced as a result of *how* you participate, what Bourdieu and Darbel called your mode of participation, not just *what* you participate in (1991). Thus, participation across practices and fields, from the elite to the popular, also marks social reproduction, a form of participation that has been described as cultural omnivorousness (Bennett et al., 2009). From this perspective, being able to participate in, accrue, use and exchange forms of capital across different practices and fields is particularly advantageous. Sarah Thornton (1996, p. 101) called this the "classless autonomy" of those able to dip in and out of cultural practices across the high-brow to low-brow spectrum and noted that this omnivorousness is in fact extremely privileged. Only those with privileged backgrounds are able to move so seamlessly between fields and practices. Drawing on the research discussed in the previous section, some people do seem to be omnivores in the case of everyday science learning. In terms of everyday science learning users, audiences or visitors being an omnivore means visiting science centres, watching science documentaries at home, taking part in a government consultation and going to a science comedy event would work together to reproduce your privilege. Thus, while everyone is involved in a variety of fields, not all forms of capital or practices are equally accessible, valued or legitimated, depending on the field in question and the relative status of an individual (Bourdieu & Wacquant, 1992; Skeggs, 2004).

Mapping (non)participation in everyday science learning practices

In this section I turn to explore participants' perspectives on everyday science learning and, specifically, their involvement (or not) in various everyday science learning practices. In other words, from their perspective and given the experiences, how accessible were the fields associated with everyday science learning? What kinds of everyday science learning practices did participants recognise or value? Which practices were they more familiar with and which were wholly invisible?

Participants' accounts of everyday science learning described a set of fields and related practices that were by and large inaccessible. From television watching to museum visiting, participants' involvement in the various fields of everyday science learning was narrow, and limited to the more popular, low-brow or mundane modes of practice, rather than special, dominant or high-brow practices. Thus, even when participants were involved in everyday science learning, these practices were those that were least likely to support the building, use of or exchange of dominant forms of capital.

Special, dominant practices

Museums were the most recognised of all the everyday science learning fields and practices that we talked about. For practitioners working in science museums this finding is notable; these institutions were the most visible forms of everyday science learning practice outside the formal school system as far as participants were concerned. This did not mean, however, that museums were accessible; far from it. In an interview, for example, Hawa and Lucille from the Sierra Leonean group, bluntly stated:

HAWA: like that those of us from Sierra Leone [. . .] maybe they need to, like science and the museums, they need to advertise them, more broadly. You see the type of people coming in?
LUCILLE: That goes there? It's not us.

This point was re-iterated frequently across all five groups. Indeed, Kadiatu and Fatima, also from the Sierra Leonean group, argued that science museums were of little interest or relevance to their community in no uncertain terms during a focus group, when I asked if they had ever or would ever consider visiting one:

KADIATU: Not science museums, no (laughs).
FATIMATA: I would go to the cinema, I would never tell anyone I would go to a museum.
KADIATU: I would say half of the Sierra Leone community they never just sit down and say, "let's go to the science museum".

Participants recognised traditional everyday science learning practices and institutions, such as museums, as a form of high-brow culture and, as a result, broadly unappealing and inaccessible for people like them (for reasons discussed more in Chapters Six and Seven). This finding is particularly problematic from an equity perspective since Bourdieu (1984) argued that the more dominant a practice, the more valuable the cultural capital involved in such practices. Visiting science museums, however, was laughable for participants.

Notably, while some participants or their children visited science museums on school visits, these were not fondly remembered, nor were these experiences that catalysed further visits. As Fatima from the Somali group argued, "I probably wouldn't go back to museums any time soon because I was taken there by force [with school]". Importantly, this point contradicts claims elsewhere in the museum studies literature that school visits to museums facilitate the inclusion of broader publics (see for example Hooper-Greenhill, Phillips, & Woodham, 2009). School groups do make up a significant proportion of visitors to science museums (and science centres and, for that matter, art galleries and social history museums), and doubtless these links to the formal school system help to make the museum field more visible (Falk et al., 2015). That these visits are necessarily positive however, or drive return visits from students and their families, is evidently not something that should be assumed. Thus, despite living in central London and being aware of London's more prestigious everyday science learning institutions, neither participants nor their children used these resources.

Newer, less traditional everyday science learning practices were invisible and, as a result, inaccessible for participants. Participants had not heard of science talks in cafés or pubs, science festivals, citizen science practices, science storytelling or science discussion events. For example, Khalid, from the Somali group, explained he would not know where to start to find such events, would not know where they took place, what they were or why he should attend. His perspective was widely echoed across all five groups. Thus, despite the proliferation of a wide range of everyday science learning practices through significant funding in the UK over recent years, such activities remained invisible to participants. Against these perspectives, claims about the potential for new or hybrid forms of everyday science learning reaching different audiences from their more traditional counterparts seem questionable (Bultitude, 2014; Kaiser, Durant, Levenson, Wiehe, & Linett, 2013).

Similarly, participants had no experience of politically oriented everyday science learning activities such as local or national government consultations on socio-scientific issues, Citizens Juries consensus conferences or other practices designed to support deliberative democracy. That is, unlike voting, these activities were not seen as traditional, visible political processes. Although participants saw specific socio-scientific issues as part of their lives (such as agriculture or climate change), participants across all five groups struggled to imagine how or why they would influence political decisions on such issues. As Kirin from the Asian group put it, despite her worries about the links between pesticides, preservatives and health, she felt unable to influence government decisions: "how to approach government [. . .] some government meeting, some bosses, government, I don't know who they are". Here, we begin to see how cultural capital, powerlessness and exclusion work together. Without the institutional knowledge of how to influence science-related decision-making or knowledge of the various consultative practices that exist, Kirin and the other participants felt powerless to influence change.

Mundane, popular practices

Turning to the fields of everyday science learning that participants did use highlights again how their practices were marked by orthodox distinctions between popular and elite forms of culture (Bourdieu, 1984; Miles & Gibson, 2016). Everyday science learning practices perceived as popular, low-brow or commonplace, particularly those in the field of mass media such as television watching and using the Internet, were familiar to participants. Participants did not however, initially describe such practices as everyday science learning, because they seemed so mundane.

Stories about science frequently appear in popular culture (Chambers, 2016; Kirby, 2011; Weitkamp, 2015). Regularly watched television programmes about science ranged from blue-chip nature documentaries to shows where science formed part of the fabric of the programme but was not the main focus, such as the comedy series *The Big Bang Theory* or the detective series *Crime Scene Investigation* (*CSI*). Like museums, practices such as watching television were highly visible, once they were framed as a form of everyday science learning. Unlike museums, however, these popular practices were woven into participants' everyday lives. As Thomas from the Sierra Leonean group put it:

THOMAS: Going to a museum is like, when last did I go to a museum? When last did any of my friends go to a museum? [. . .] There's so much on the Internet and TV, and like if someone says something to do with science to you, I can just go on YouTube, and write it in and have a look, or I can watch a show like, what's that guy's name, he did that series *Life*?

EMILY: Attenborough?

THOMAS: Yeah, David Attenborough, like I found that wicked, that *Life* series, I think a lot of people must have watched that show.

Thus, in line with Bourdieu and Johnson's (1993) work on restricted access and class, popular, daily practices, such as watching television, were seen as more relevant and more accessible by participants than museums.

To put this in context, watching television appears to be a ubiquitous cultural practice in the UK, even for people largely excluded from other forms of dominant cultural participation (Bennett et al., 2009; Taylor, 2016). Notably, while television and the Internet were identified as everyday science learning practices that participants regularly encountered, few participants sought out science through these media. Rather, daily cultural practices around television watching and going online sometimes overlapped with science content. For instance, as Irene from the Afro-Caribbean group put it:

IRENE: I don't know much about science (laughs).

EMILY: You say you enjoyed listening to it.

IRENE: On the telly I enjoy listening to some things, but it's not something that really attracts me.

In other words, although Irene did encounter science through watching television, like many of the other participants, that did not mean she actively pursued science content on television.

Watching science-related television programmes did not necessarily provide participants with science capital, especially in terms of whether it supported them to identify with science or feel like science was for them (Archer et al., 2015). Watched for entertainment value, science on television featured an all-star cast of people who were, in Irene's words, "not like us". As Kirin from the Asian group put it when talking about the television series *CSI*, "we're very interested but you know, we can't push ourselves forward". For her, her friends and participants from other groups, science on television was represented by people who were special, impossibly clever, but not like them and did things they could not do, contributing to their sense that science was difficult and the reserve of what Kirin described as "geniuses" (Lemke, 1990). In the same breath as participants named a series of famous white male science presenters, including Sir David Attenborough, Steve Irwin and Sir Patrick Moore, they highlighted the differences they saw between themselves and their perception of the kinds of people who were involved with science.

With a few exceptions, like Fatima, whose story is described in more detail in Chapter Four, for most participants watching science programmes on television did not result from a pre-existing interest in science, nor did it particularly contribute to developing such interests. Watching television represented a more accessible mode of everyday science learning, woven as it was into the fabric of participants' daily lives and integrated into their homes. But it did not seem to reduce the distance between participants and both science and everyday science learning. This finding raises questions about access across the different fields of everyday science learning and which forms of capital are available for participants whose involvement in everyday science learning was limited to television and the Internet (Bourdieu, 1984). Such questions are important not least because high/low patterns of cultural value still operate within these fields. Thus, a so-called blue chip science documentary from the BBC is more prestigious than comedy show with science as a background context (Bennett, 2006; Dingwall & Aldridge, 2006; Li & Orthia, 2016). In other words, can watching *CSI* on television provide the same advantages as taking part in a broader range of everyday science learning activities?

Restricted fields of everyday science learning

Mapping participants' involvement in and recognition of everyday science learning practices shows that participation was narrow (limited to science media consumption) and patterned by what have been understood as classed distinctions about elite/popular practices (Bourdieu, 1984; Miles & Gibson, 2016; Taylor, 2016). From this data alone it is not evident how racialised these patterns might be. However, Trienekens (2002) and Erel (2010) have argued in multi-ethnic societies, 'race'/ethnicity accounts more for patterns of cultural consumption than class alone.

Similarly, research on art galleries, ethnographic, social history and science museums in different countries has found 'race'/ethnicity, racism and colonialism significantly influence practices in these sites and, as a result, visitors' experiences (Dixon, 2016; Garibay, 2017; Hahn, 2016). Thus, not only are museum content and representation racialised, but so too are the socio-political histories of those institutions, their staff and visitors. Furthermore, as discussed in subsequent chapters, participants detailed accounts of the experiences of every-day science learning highlight more clearly how 'race'/ethnicity and racism influenced their participation.

Participants' recognition of and involvement in practices they saw as everyday science learning reflected the official or orthodox perspective on cultural activities that Miles and Gibson argue operates in the UK context (2016). That is, the more elite or dominant a practice, the more recognisable it was, and the more restricted and inaccessible that field was for participants. In contrast, low-brow, popular activities, such as watching television, were more accessible but not seen as particularly important forms of everyday science learning by participants, rather they were seen as mundane.

Participants' involvement in everyday science learning practices was narrow, rather than broad. That is, they were not everyday science learning 'omnivores', nimbly picking and choosing across the various fields of everyday science learning and the practices within them (Bennett et al., 2009). Instead, they were limited to those practices that they could access (popular cultural activities such as watching science-related television) and, as a result, were unable to access and accrue certain, more powerful forms of capital related to other fields of everyday science learning. Participants' experiences therefore suggest the fields involved in everyday science learning mirror long-established patterns of classed and racialised exclusion from high-brow culture, education and politics (Bennett et al., 2009; Bourdieu & Darbel, 1991; Bourdieu & Johnson, 1993; Bourdieu & Passeron, 1990; Miles & Gibson, 2016).

Indeed, for Bourdieu, cultural practices become legitimate and dominant through state support and visibility (even to those unable or unwilling to participate) combined with restricted access (1984; Bourdieu & Johnson, 1993). Thus, the more elite a practice, the more visible it is and the more likely that access to such a practice is restricted. Participants' views and experiences of everyday science learning clearly echo this pattern. In other words, the more dominant a given everyday science learning field or practice, the more it was marked by inaccessibility for participants. Thus, from a cultural consumption perspective, the fields and practices involved in everyday science learning were significantly marked by inequality, inaccessibility and disadvantage for participants. As such, participants experiences confirm the partial patterns of participation suggested by the larger data sets described earlier in this chapter; the way everyday science learning resources are used, advantages people from dominant social groups at the expense of people from minoritised groups.

Mapping participants' cultural consumption beyond science

One pitfall to beware with any analysis of cultural consumption is how to negotiate the question of participatory deficit in relation to culture, education and politics. To provide more context against which to understand participants' experiences of exclusion and everyday science learning, in this section I discuss the kinds of cultural, educational and political activities that participants *were* involved in. As such, my point is to illustrate how partial a deficit approach to participation in everyday science learning can be, and quite how much a deficit approach might overlook. The people who took part in the research presented in this book were culturally, educationally and politically active. As such it is not appropriate or accurate to frame their non-participation in everyday science learning practices as rooted in a participatory deficit on their part.

Ideas about social inclusion shifted in the 1990s in the UK, away from a focus on socio-economic factors such as access to job markets, to focus instead on access to cultural practices (broadly defined here as those cultural, educational and/or political practices that are public and, in at least some senses, claim to be open for people to participate in) (Levitas, 1998, 2004; Sandell, 1998). Against such a shift, it is important to think carefully about which cultural practices count and for whom they matter. As arguments in the education research literatures suggest, valuing the practices, knowledges and assets people are familiar with fundamentally reorients how we understand the experiences of minoritised groups, away from a focus on what they don't, can't or won't do (Barton & Tan, 2009; Gonzalez, Moll, & Amanti, 2013; Ladson-Billings, 2005). Everyday science learning practices were evidently not priorities in participants' lives. In contrast, participants across all five groups were heavily involved in community-based cultural, educational and political activities. Although they carry different nuances, I discuss these practices together here to provide a backdrop against which their non-participation in and exclusion from everyday science learning can be reframed.

Over the two years I carried out fieldwork for this project I met participants most regularly in their community centres, but we also met at festivals in those centres, in parks, in other community spaces and participants' homes. It was obvious that participants led rich cultural lives. For instance, when I turned up one sunny April morning to interview three members of the Asian group at their community centre I was met with sticky, sweet jalebi[2] (a popular Indian dessert made of fried batter soaked in sugar syrup) and was invited to join their afternoon of playing music, singing and dancing. This was not uncommon during fieldwork.

Indeed, the combination of food, speaking 'home' languages, dance and music was something all participants valued and seemed to be almost constantly involved with. All of these practices can be understood as forms of community-based cultural capital (Trienekens, 2002). Thus meetings at the Sierra Leonean community centre regularly involved a celebration of one kind or another, with everyone bringing food. Enough of these events involved dancing that the Sierra Leonean elders mercilessly mocked me for my two left feet[3] while they, in their 70s and 80s, danced elegantly. During the fieldwork research participants were involved in community-based cultural activities on an approximately weekly basis, and, for many participants this was a daily practice (for instance, in the case of those practicing music or languages), while larger community festivals involved months of preparation.

The cultural practices important to participants and their communities were seamlessly woven into their lives, unlike everyday science learning practices. As Erel (2010) has argued, racialised groups living in diaspora work hard to re-create culturally valued practices in ways that draw on hybrid forms of 'home' and 'host' cultures. Participants were almost constantly involved in such practices, often in ways that involved a significant effort on their part, significant efforts that they took for granted. Elena, for instance, a Mexican participant from the Latin American group, explained to me one day about how much work she did organising cultural events as well as performing at them:

ELENA: I help organise. In the project I'm the one that gets all the money together, and I sing and play guitar as well, I also dance. I also organise the group workshop side of it.

EMILY: So how much time does that take you?

ELENA: A lot, I have to take, we have to practice and at the moment we practice twice a week for the folk group as well.

EMILY: And can I ask you as well, do you get paid for it?

ELENA: Do I get paid for it, no (laughs) no.

EMILY: And do you have space to do it in?

ELENA: At the moment we practice in the basement of a restaurant, which they kindly let us use, yeah. I like to organise the Mexican events, we try to always get a Mexican point to the party, and this year will be the biggest one because we're celebrating 200 years of independence in Mexico.

As this quote from an interview with Elena shows, participants were committed organisers of community-based cultural events. She did not get paid for her efforts. She practiced with her band twice a week. She organised the accounts on behalf of the group of people producing the celebration of Mexican independence. She also held down a full-time job in a local, community-owned business as well as working ad hoc cleaning shifts for various offices in central London.

To understand the full remit of participants activities it is helpful to extend the notion of community-based cultural practices to include community-based educational and political practices (Trienekens, 2002). Although I have described a focus across all five groups on food, language, music and dance, I do not mean to obscure the huge range of activity participants were involved in within their communities. In terms of education, almost every group[4] ran language classes for adults and children as well as art and music workshops. The Somali group ran sports activities for their children and participants from the Latin American group were passionate supporters of their own football league.

Participants were involved in health and housing projects as well as local political campaigns. The Sierra Leonean group organised meetings to discuss community concerns ranging from family breakdown to supporting more Sierra

Leonean youth to stay engaged in school to lobbying local government for better representation of Sierra Leoneans in civic data. Participants from the Somali group were active in keeping up pressure on the Home Office regarding the refugee status of particular community members. Similarly, participants from the Latin American group were involved in a London-wide campaign to reject the newly introduced practice amongst their various embassies and London Borough Councils to describe their community in London as 'Ibero-American', which they found extremely offensive. Like the Sierra Leonean participants, Latin American participants were concerned about their political visibility, the position and safety of 'illegal' migrants and their civil rights (see also Pero, 2010).

For participants across all five groups, the community-based cultural activities that they produced, performed and enjoyed within the sphere of their own communities were an extremely important part of their lives, often taking up what little time was available between work and family life. As Maria, a Colombian participant from the Latin American group, put it:

MARIA: Most people are pretty knackered all the time. And then, the only thing is, if they get to go out, say on a Saturday night, to one of the Latin places and dance all night 'til four in the morning, that's how they unwind, the only way. Or there are a lot of people playing football.
EMILY: 'Cos there's always people on those football pitches.
MARIA: Exactly, and singing and dancing and football, that's the main thing.

To understand the role played by these community-based cultural activities in people's lives it is important to remember that as a result of their migration, participants occupied multiple social positions, simultaneously, in different countries. For example, although resident in the UK, all participants maintained strong links with family members, friends and colleagues in their 'home' countries as well as in several other countries depending on their community diaspora. In this sense, for participants, 'race'/ethnicity was constructed in ways that could not be narrowly defined as relating to either a 'home' country or as a minority group in a 'host' country. Instead it is better understood as transnational, as a result of migration trajectories, colonialism and globalisation (Bhabha, 1994; Hall, 2017; Vertovec, 2004).

Each group included participants whose personal migration trajectories involved living in several countries. Amongst the Asian group, for instance, India was described as one home amongst several others, albeit a home that some participants had never lived in. Roger Rouse and Steven Vertovec (1992, p. 41, 2004, p. 974) have referred to this as "bifocality" of perspective, arguing that migrants develop bifocality as part of adjusting to the dual, if not multiple sets of references required of living in more than one country.

Away from migration studies these ideas have also been discussed in terms of post-colonial theories of belonging, displacement, 'race'/ethnicity and culture (Bhabha, 1994, 1996; Gilroy, 1993; Modood, 2010; Spivak, 1999). My point here is simply to note that while bifocality can be linked to migration, it is not divorced from history. In trying to navigate a transnational sense of self, participants were involved in careful balancing work to maintain a dual, or multiple, sense of belonging. It was this multiplicity of perspective that made participation in community-based cultural activities so important to participants. This sense of self was not simply a matter of migration, but was tied up with socio-political and historic patterns of colonialism and contemporary globalisation in ways that echoed through the generations. Participants needed spaces and practices that validated what they held dear, especially in the face of their otherwise apparent cultural, educational or political invisibility in the UK. Drawing on the work of Homi K. Bhabha (1996, p. 55), participants' community-based cultural capital practices can be understood as ways to both assert and protect "culture-as-difference". Such practices were valuable precisely because they maintained and re-created cultural practices as assets for participants.

Of course, participants were recruited through their involvement in grass-roots community groups. That means we should not assume their level of involvement in community-based cultural, educational or political activities is generalisable across other people from similar ethnic or socio-economic background necessarily. What these data show however, are the limits of a deficit approach to understanding participants in relation to everyday science learning. Taking a deficit approach makes so much of participants lives and their practices, knowledges and values invisible.

Participants were clearly far from deficient in terms of their participation in cultural, educational and/or political activities. Their extensive involvement in community-based cultural practices (broadly defined) is interesting for this book precisely in comparison to their limited and narrow involvement in everyday science learning. The relative differences between these two patterns of participation clearly speak to issues of relevance and accessibility in terms of cultural consumption.

Community-based cultural events were organised such that they were often free or very cheap to attend. They were local (typically in participants' home neighbourhoods and within walking distance). Community-based cultural activities were produced by participants for themselves. These practices were enmeshed in participants' everyday lives and were deeply meaningful. In contrast, everyday science learning activities were less accessible, less relevant and, unlike community-based cultural practices, perceived by participants as I discuss in more detail in the next four chapters, as "not for us". Thus, as I argued in Chapter Two, instead of understanding exclusion in terms of a deficit on the part of the excluded, we need to think about how everyday science learning practices are exclusive, off-putting and/or seem less relevant than community-based practices.

Summary

This chapter explored participants' involvement in everyday science learning and their views of such practices from a cultural consumption perspective. In attending to participants' experiences, I have tried to provide an alternative starting point for how we might think about exclusion and, by extension inclusion, as discussed in detail in Chapter Seven (Ortega, 2006). The analysis presented here highlights the need to leave behind arguments about participatory, cultural or political deficits when imagining people – or publics – who do not or cannot participate in everyday science learning (Levitas, 2004; Miles & Gibson, 2016). First, because mapping participants involvement in everyday science learning suggests these are restricted fields of cultural production, that are marked by and reproduce structural inequalities. And second, because participants were far from deficient in terms of their participation, but the practices they valued were rarely recognised or respected beyond their communities.

The data discussed at the start of this chapter suggest that participation in everyday science learning plays a role in social reproduction because exclusion is hierarchical (structured by social advantage and disadvantage) and intersectional (since multiple structural inequalities are involved). Participants' experiences build on this picture to suggest that the fields involved in everyday science learning act, just as Bourdieu argued in relation to social reproduction, via arts, education and cultural participation, to preserve advantages for the privileged at the expense of minoritised groups (Bourdieu, 1984; Bourdieu & Darbel, 1991; Bourdieu & Johnson, 1993). Thus, the more dominant a field and the more valuable the capital within it, the more inaccessible it was for participants. At the same time, the practices and knowledges participants valued were embedded in their lives, but not represented, from their perspective, in everyday science learning.

The data discussed in this chapter also suggest that constructions of the public sphere that rely on dualistic, gendered and Eurocentric assumptions about which forms of participation count (public/private, everyday/special) need to be reimagined (Ebrey, 2016; Fraser, 1990). Thus, building on research about funds of knowledge, focusing on people's *assets* rather than their perceived *deficits*, seems likely to be a better starting point for understanding what inclusive everyday science learning might involve, as discussed in more detail in Chapter Seven (Gonzalez et al., 2013). We should be careful, however, not to forget how deeply enmeshed racialised power dynamics are in which forms of capital are valued (Ladson-Billings, 2005). As this chapter has shown, not all forms of cultural capital are valued in the same ways. Nonetheless, taking into account a broader sense of 'what counts' might help to reimagine 'who counts' in more inclusive terms.

Notes

1 The *Mendoza Review* (Mendoza, 2017) and *Strategic Review of DCMS sponsored Museums* (Department for Digital Culture Media and Sport, 2017) lists those museums currently receiving significant support through direct government funding in the UK. Under the New Labour government (1997–2010), several museums across the UK were 'nationalised', meaning their entry fees were removed via government subsidy.

2 Participants in every group always made sure I was well fed. Because I did the fieldwork as part of my PhD participants frequently asked about my well-being and how "broke" I was. Although I always brought cakes or biscuits to community events, I was also the happy recipient of boxes of food to take home at the end of such events on more than one occasion. I make this point, perhaps unnecessarily, because I want to highlight that the ethics of care in this project went both ways, and I remain deeply grateful to participants for more reasons than their involvement in the data generated through this research.

3 To be honest, participants in every group mocked my dancing; it was not just the Sierra Leonean elders. To me this seemed to be part of the 'in-betweenness' of fieldwork. Participants and I together echoed certain cultural expectations about who could and could not dance based on our cultural backgrounds and bodies. Thus, I was the white girl, and a science insider, and someone who was not very good at dancing.

4 The Afro-Caribbean group did not run a language class; the four other groups did.

References

Archer, L., Dawson, E., DeWitt, J., Seakins, A., & Wong, B. (2015). "Science capital": A conceptual, methodological, and empirical argument for extending bourdieusian notions of capital beyond the arts. *Journal of Research in Science Teaching*, 52, 922–948.

The Association for Science and Discovery Centres. (2010). *Assessing the impact of UK science and discovery centres: Towards a set of common indicators*. Bristol: Association for Science & Discovery Centres.

Atkinson, R., Siddall, K., & Mason, C. (2014). *Experiments in engagement: Engaging with young people from disadvantaged backgrounds*. London: Wellcome Trust.

Barton, A. C., & Tan, E. (2009). Funds of knowledge and discourses and hybrid space. *Journal of Research in Science Teaching*, 46(1), 50–73.

Bell, P., Lewenstein, B., Shouse, A. W., & Feder, M. A. (2009). *Learning science in informal environments: People, places, and pursuits*. Washington, DC: The National Academies Press.

Bennett, T. (2006). Distinction on the box: Cultural capital and the social space of broadcasting. *Cultural Trends*, 15(2), 193–212.

Bennett, T., Savage, M., Silva, E., Warde, A., Gayo-Cal, M., & Wright, D. (2009). *Culture, class, distinction*. Abingdon and New York: Routledge.

Bhabha, H. K. (1994). *The location of culture*. Abingdon and New York: Routledge.

Bhabha, H. K. (1996). Culture's in-between. In S. Hall & P. Du Gay (Eds.), *Questions of cultural identity* (pp. 23–60). London, Thousand Oaks, CA and New Delhi: Sage.

Bourdieu, P. (1984). *Distinction: A social critique of the judgement of taste* (R. Nice, Trans.). London: Routledge.

Bourdieu, P., & Darbel, A. (1991). *The love of art: European art museums and their public*. Oxford: Polity Press.

Bourdieu, P., & Johnson, R. (1993). *The field of cultural production: Essays on art and literature*. Cambridge: Polity Press.

Bourdieu, P., & Passeron, J.-C. (1990). *Reproduction in education, society and culture* (R. Nice, Trans., 2nd ed.). London, Newbury Park, CA and New Delhi: Sage.

Bourdieu, P., & Wacquant, L. (1992). *An invitation to reflexive sociology*. Chicago: University of Chicago Press.

Broks, P. (2006). *Understanding popular science*. Maidenhead: Open University Press.

Bultitude, K. (2014). Science festivals: Do they suceed in reaching beyond the "already engaged"? *Journal of Science Communication, 13*(4).

Bultitude, K., McDonald, D., & Custead, S. (2011). The rise and rise of science festivals: An international review of organised events to celebrate science. *International Journal of Science Education, Part B, 1*(2), 165–188. doi.org/10.1080/2154845 5.2011.588851

Chambers, A. C. (2016). The evolution of *Planet of the Apes*: Science, religion, and 1960s cinema. *The Journal of Religion and Popular Culture, 28*(2–3), 107–122. doi:10.3138/jrpc.28.2-3.3399

Davies, S. R., & Horst, M. (2016). *Science communication: Culture, identity and citizenship*. London: Palgrave Macmillan.

Dawson, E. (2014). Equity in informal science education: Developing an access and equity framework for science museums and science centres. *Studies in Science Education, 50*(2), 209–247. doi.org/10.1080/03057267.2014.957558

Delgado, A., Lein Kjølberg, K., & Wickson, F. (2011). Public engagement coming of age: From theory to practice in STS encounters with nanotechnology. *Public Understanding of Science, 20*(6), 826–845.

Department for Culture Media and Sport. (2011). *Taking part: The national survey of culture, leisure and sport*. London: Department for Culture, Media & Sport.

Department for Culture Media and Sport. (2016). *Taking part: Longitudinal report 2016*. London: Department for Culture, Media & Sport.

Department for Digital Culture Media and Sport. (2017). *Strategic review of DCMS sponsored museums*. London: Department for Culture, Media & Sport.

DeWitt, J., & Archer, L. (2017). Participation in informal science learning experiences: The rich get richer? *International Journal of Science Education, Part B, 7*(4), 356–373. doi:10.1080/21548455.2017.1360531

Dingwall, R., & Aldridge, M. (2006). Television wildlife programming as a source of popular scientific information: A case study of evolution. *Public Understanding of Science, 15*(2), 131–152. doi:10.1177/0963662506060588

Dixon, C. A. (2016). *The "othering" of Africa and its disaporas in Western museum practices*. (PhD monograph), University of Sheffield, Sheffield.

Duensing, S. (2006). Culture matters: Informal science centes and cultural contexts. In Z. Bekerman, N. C. Burbules, & D. Silberman Keller (Eds.), *Learning in places: The informal education reader* (pp. 183–202). New York: Peter Lang Publishing.

Ebrey, J. (2016). The mundane and insignificant, the ordinary and the extraordinary: Understanding everyday participation and theories of everyday life. *Cultural Trends, 25*(3), 158–168. doi.org/10.1080/09548963.2016.1204044

Ecsite-UK. (2008). *The impact of science and discovery centres: A review of worldwide studies*. Bristol: Association for Science & Discovery Centres.

Erel, U. (2010). Migrating cultural capital: Bourdieu in migration studies *Sociology, 44*(4), 642–660.

Falk, J., Dierking, L. D., Osborne, J., Wenger, M., Dawson, E., & Wong, B. (2015). Analyzing science education in the United Kingdom: Taking a system-wide approach. *Science Education, 99*(1), 145–173.

Feinstein, N. W. (2017). Equity and the meaning of science learning: A defining challenge for science museums. *Science Education, 101*(4), 533–538. doi:10.1002/sce.21287

Fraser, N. (1990). Rethinking the public sphere: A contribution to the critique of actually existing democracy. *Social Text*, (25/26), 56–80.

Garibay, C. (2017). *Metasynthesis of front-end studies excerpt, Winter 2017*. Chicago: Garibay Group.

Gillborn, D. (2010). The colour of numbers: Surveys, statistics and deficit-thinking about race and class. *Journal of Education Policy*, 25(2), 253–276.

Gillborn, D., Warmington, P., & Demack, S. (2018). QuantCrit: Education, policy, "Big Data" and principles for a critical race theory of statistics. *Race Ethnicity and Education*, 21(2), 158–179. doi:10.1080/13613324.2017.1377417

Gilroy, P. (1993). *The Black Atlantic: Modernity and double consciousness*. London: Verso.

Gonzalez, N., Moll, L. C., & Amanti, C. (2013). *Funds of knowledge: Theorizing practices in households, communities, and classrooms*. Mahwah, NJ: Taylor & Francis.

Hahn, C. N. (2016). *The political house of art: The South African National Gallery 1930–2009*. (PhD Monograph), Goldsmiths College, University of London, London.

Hall, S. (2017). *The fateful triangle: Race, ethnicity, nation* (K. Mercer, Ed.). Cambridge, MA and London: Harvard University Press.

Hooper-Greenhill, E., Phillips, M., & Woodham, A. (2009). Museums, schools and geographies of cultural value. *Cultural Trends*, 18(2), 149–183.

Ipsos MORI. (2001). *Visitors to museums and galleries in the UK*. London: Ipsos MORI.

Ipsos MORI. (2006). *Renaissance in the regions 2005: Visitor exit survey: Final national report*. London: Museums Libraries Archives.

Ipsos MORI. (2011). *Public attitudes to science 2011*. London: Department for Business, Innovation & Skills.

Ipsos MORI. (2014). *Public attitudes to science 2014*. London: Department for Business, Innovation & Skills.

Ipsos MORI. (2016). *Wellcome trust monitor report wave 3: Tracking public views on science and biomedical research*. London: Wellcome Trust.

Kaiser, D., Durant, J., Levenson, T., Wiehe, B., & Linett, P. (2013). *The evolving culture of science engagement*. Cambridge, MA: MIT & Culture Kettle.

Kirby, D. (2011). *Lab coats in Hollywood*. Cambridge, MA and London: MIT Press.

Ladson-Billings, G. (2005). The evolving role of critical race theory in educational scholarship. *Race Ethnicity and Education*, 8(1), 115–119. doi:10.1080/136133 2052000341024

Lareau, A., & Weininger, E. B. (2003). Cultural capital in educational research: A critical assessment. *Theory and Society*, 32(5–6), 567–606. doi:10.1023/B:RYSO. 0000004951.04408.b0

Lemke, J. L. (1990). *Talking science: Language, learning, and values*. Westport: Ablex Pub. Corp.

Leong, N. (2013). Racial capitalism. *Harvard Law Review*, 126(8), 2153–2226.

Levitas, R. (1998). *The inclusive society?* Basingstoke and New York: Palgrave Macmillan.

Levitas, R. (2004). Let's hear it for Humpty: Social exclusion, the third way and cultural capital. *Cultural Trends*, 13(2), 41–56.

Li, R., & Orthia, L. A. (2016). Communicating the nature of science through the Big Bang Theory: Evidence from a focus group study. *International Journal of Science Education, Part B*, 6(2), 115–136. doi:10.1080/21548455.2015.1020906

Lorde, A. (1984). *Sister outsider*. Berkeley: Crossing Press.

Massarani, L., & Merzagora, M. (2014). Socially inclusive science communication. *Journal of Science Communication, 13*(2), 1–2.

McGuigan, J. (1996). *Culture and the public sphere*. London and New York: Routledge.

Mendoza, N. (2017). *The Mendoza Review: An independent review of museums in England*. London: Department for Digital, Culture, Media & Sport.

Miles, A., & Gibson, L. (2016). Everyday participation and cultural value. *Cultural Trends, 25*(3), 151–157. doi.org/10.1080/09548963.2016.1204043

Modood, T. (2010). *Still not easy being British: Struggles for a multicultural citizenship*. Stoke on Trent: Trentham Books.

National Science Foundation. (2012). *Science and engineering indicators 2012*. Arlington, VA: National Science Foundation.

OECD. (2012). *Education at a glance 2012: OECD indicators*. Retrieved from www.oecd-ilibrary.org/education/education-at-a-glance-2012_eag-2012-en

Ortega, M. (2006). Being lovingly, knowingly ignorant: White feminism and women of color. *Hypatia, 21*(3), 56–74. doi:10.1111/j.1527-2001.2006.tb01113.x

Pero, D. (2010). Political engagement and Latin Americans in the UK. In W. Lem & P. Garinder Barber (Eds.), *Class, contention and a world in motion* (pp. 81–104). New York and Oxford: Berghahn Books.

Prieur, A., & Savage, M. (2013). Emerging forms of cultural capital. *European Societies, 15*(2), 246–267. doi.org/10.1080/14616696.2012.748930

Puwar, N. (2004). *Space invaders: Race, gender and bodies out of place*. Oxford and New York: Berg.

Randhawa, K., & Spanier, G. (2014). Hawking hails London, capital of science. *London Evening Standard*, p. 1.

Rouse, R. (1992). Making sense of settlement: Class transformation, cultural struggle, and transnationalism among Mexican migrants in the United States. *Annals of the New York Academy of Sciences, 645*(1), 25–52. doi:10.1111/j.1749-6632.1992.tb33485.x

Saha, A. (2018). *Race and the cultural industries*. Cambridge: Polity Press.

Sandell, R. (1998). Museums as agents of social inclusion. *Museum Management and Curatorship, 17*(4), 401–418.

Scott, S. (2009). *Making sense of everyday life*. Cambridge: Polity Press.

Skeggs, B. (2004). *Class, self, culture*. London and New York: Routledge.

Spivak, G. C. (1999). *A critique of postcolonial reason: Toward a history of the vanishing present*. Cambridge, MA and London: Harvard University Press.

Stilgoe, J., Lock, S. J., & Wilsdon, J. (2014). Why should we promote public engagement with science? *Public Understanding of Science, 23*(1), 4–15.

Taylor, M. (2016). Taking part: The next five years (2016) by the Department for Culture, Media and Sport. *Cultural Trends, 25*(4), 291–294.

Thornton, S. (1996). *Club cultures: Music, media, and subcultural capital*. Hanover, NH: University Press of New England.

Thorpe, C., & Gregory, J. (2010). Producing the post-fordist public: The political economy of public engagement with science. *Science as Culture, 19*(3), 273–301.

Trienekens, S. (2002). "Colourful" distinction: The role of ethnicity and ethnic orientation in cultural consumption. *Poetics, 30*(4), 281–298.

Vertovec, S. (2004). Migrant transnationalism and modes of transformation. *International Migration Review, 38*(3), 970–1001. doi.org/10.1111/j.1747-7379.2004.tb00226.x

Weitkamp, E. (2015). A question of (audience) reach. *Journal of Science Communication, 14*(3), 1–4.

Wellcome Trust. (2008). *Millennium science centres impact assessment report: Executive summary.* London: Wellcome Trust.

Wiehe, B. (2014). When science makes us who we are: Known and speculative impacts of science festivals. *Journal of Science Communication, 13*(4).

Chapter 4

No 'taste' for science?

In this chapter I explore the science part of the equity, exclusion and everyday science learning puzzle. Everyday science learning is strongly framed around its content – science. That science is enmeshed in socio-political histories of power that influence contemporary life is beyond dispute (Agar, 2012; Ravetz, 2005). Though science can sometimes seem like an a-political enterprise, experience and research teaches us otherwise. Science has been used to exacerbate structural inequalities, not least through scientific research on 'race'/ethnicity, class and gender, that has essentialised and demonised particular people at particular times (see for example, Hammonds, 2009; Haraway, 1997; Medin & Bang, 2014; Tall-Bear, 2013). Nothing about science is neutral.

The word 'science' encompasses a vast array of different subjects, practices and people, with complex and varying relationships between those sciences, technologies and communities (Knorr Cetina & Mulkay, 1983; Medin & Bang, 2014; Michael, 2006; Ziman, 2002). Science content can be understood in many ways, for instance, as scientific knowledge, practices and skills, as people and the scientific community or as technologies and other applications of science. Nonetheless, for all its problems and complexity, science remains the key signifier of everyday science learning. As a result, examining what people think and feel about science as well as their previous experiences of science is useful because these are lasting and salient features of how people relate (or not) to everyday science learning (Osborne, Simon, & Collins, 2003).

I begin in the first part of this chapter by exploring participants' attitudes towards and expectations of science, drawing on theoretical tools from Bourdieu (1992) – particularly the idea of habitus – and Mike Michael's (1992, 2006) work on publics and science. In this chapter I argue that participants' attitudes towards science ought not be understood as simply anti-science or ignorance, but rather as a more nuanced set of sometimes contradictory perspectives that ultimately disposed them against science.

In the second part of this chapter I discuss the stories of three participants in more detail. These three participants stood out from their friends and families, as you might imagine after reading the first part of this chapter, because they told me they liked science. Notable across their stories are shared experiences about how structural inequalities shaped their encounters with science. These stories provide a context for the pervasive dispositions that participants across the five community groups shared against everyday science learning, even for those few who liked science. Indeed, the story of Fatima, a Somali participant in her 20s, provides a case in point for how liking science does not translate straightforwardly into taking part in everyday science learning.

Throughout this chapter I argue that exploring participants' experiences can help us to move the conversation about exclusion from science away from presumptions of deficit and ignorance. Instead, I argue that the accumulation of shared experiences of science so marked by structural inequalities, can position whole communities as outside science in ways we need to take more seriously.

Attitudes towards science

One of the founding myths of much of the research on science and society relationships is that people do not participate in everyday science learning because they do not like science, and that they do not like science because of their ignorance and negative attitudes (Lock, 2011; McNeil, 2013). This has been a powerful idea. In the UK, for instance, the whole field of science communication (practice and research) was galvanised into being by the compelling case made in the Royal Society's so-called Bodmer Report (1985) based on this very premise. That people who held negative attitudes towards science were ignorant and their mistake might be remedied through education. The idea that greater public understanding of science would equate to greater public support for science (not least financial support in the form of public subsidy) has provided the underlying rationale for institutional, organised and government funded everyday science learning in the UK. This assumption is also deeply embedded in both the deficit and crusade approaches to exclusion and inclusion, as discussed in Chapter Two. And it is this idea that accounts for the peculiar fact that much of the work on science and society relationships, in the UK at least, has sought to counter what have been perceived as ignorant and/or negative attitudes towards science though prescribing a diet of *more* exposure to science through activities such as those involved in everyday science learning.

Participants' attitudes towards and experiences of science, as I discuss in this section, were more complicated than simply disliking or being ignorant about science. In somewhat contradictory ways science was seen variously as everywhere and nowhere, wonderful and difficult, prestigious, aspirational and impossible, sometimes within the same conversation. Despite these contradictions, participants shared a sense that science and science-related activities were not relevant, useful or for them. That is, they had no taste for science. As Fatima put it, when talking about why her community avoided science-related activities, "Why do something you don't do? It's not part of you".

Michael's (1992, p. 313) work on discursive formations of science "in general" and "in particular" provides a useful way to map some of the nuances of how participants described science. Michael (1992) examined how people positioned themselves in relation to science. He argued science in general was seen as impenetrable and impossible. In contrast, science in particular was more tangible and something people could relate to and sometimes identify with. Participants in this project also configured science quite differently when it was understood in general terms versus as a familiar, mundane practice. I argue these differences have implications for how science might be configured in everyday science learning that could support efforts to develop meaningfully inclusive practice. I combine Michael's (1992) concepts with those of Bourdieu (1992), particularly the idea of habitus, to discuss how the idea of dispositions can make space for some of the more contradictory aspects of participants' attitudes towards science.

Unlike participants' critiques of everyday science learning as racialised, classed and, for some, gendered (as discussed in the next two chapters), participants rarely discussed science in these terms. That does not mean however, that structural inequalities did not influence their dispositions against science. Rather, that while participants talked about science as elitist, they rarely framed that explicitly in terms of 'race'/ethnicity, class or gender. With a few exceptions, discussed later, participants typically saw disliking science as a personal problem. It is only when we think about all their stories together that we can see how, to use a clichéd phrase, the personal is political.

Science in general

Participants' descriptions of science were configured around the notion of science in general in two main ways. Firstly in terms of science as a wondrous and ever-present part of their lives, science was part of everything around them. Secondly, because science was so wondrous, it was the exclusive domain of those who were clever enough to understand it. Both of these configurations of science followed Michael's (1992, p. 320) logic that "science-in-general is used as a means of distancing self from science".

That science was everywhere, special and deeply enmeshed in their lives, was something brought up by participants in every group. As Ibrahim from the Sierra Leonean group put it, "people understand that science plays a daily part of their everyday life, but even the water you drink, it's science, it goes through the water purification process, it's science and modern technology". Similarly, Kirin from the Asian group told me, "I realise that science is important, without science we are nowhere, medical, transport, air, wind, everything". Science seemed, in these kinds of statements, to be all-encompassing and omnipresent. This sense of wonder was, as Michael (1992) found, not necessarily something that supported participants to identify with science or to feel able to be involved in science. Thus, despite seeing science "everywhere" in the same breath Alejandro from the Latin American explained that "science . . . it's a subject very far from my reality, from what I do".

The perception that science was wondrous fed into the second key framing of science in general for participants: science as something only the "super clever" could pursue, as Abdou from the Sierra Leonean group put it. Research on science and mathematics education has repeatedly shown that these subjects are configured as difficult, complicated topics suitable for only the most clever of students (where what is understood as clever is too often framed by masculinity, upper and middle class practices and whiteness), effectively putting off or excluding students who do not fit this mould (Archer et al., 2019; Chaffee & Gupta, 2018; Dawson et al., 2019; Hodari, Ong, Ko, & Smith, 2016; Hussénius, 2014; Ong, Wright, Espinosa, & Orfield, 2011; Osborne et al., 2003). Thus the ideal subjects created through these ways of framing science are boys, typically from white and middle-class backgrounds (Archer et al., 2010; Carlone, Scott, & Lowder, 2014; Ulriksen, Madsen, & Holmegaard, 2010). Participants' views of science both reflected and resisted this exclusive, gendered, classed and racialised framing, which drew in many ways on the foundational narrative of science as special.

For those participants who had been to school, school remained a key context for how they saw themselves in relation to science long after their school days were over, and irrespective of which country they had been to school in. Science education research has found that school science has a widespread influence on how people see science, not least in the seemingly inescapable framing of science in general in terms of school subjects, namely, biology, chemistry and physics (Osborne et al., 2003). While experiences of science in formal learning settings were repeatedly invoked in participants' descriptions of science, these were not positive stories. As Maria, a Latin American participants in her mid-40s with four children in the British school system explained, "the way science is presented at school is very boring and uninspiring". Indeed, for almost all of the participants, school science became a contested space for participants, whether through their own experiences or in stories they told about friends and family.

Participants in every group described science as something they were not clever enough or knowledgeable enough to be involved with. As Mirza, a woman in her 40s from the Asian group put it, "I am not a science person, believe me", adding a few minutes later:

> science is for people who want to be doctors, do biology, those sorts of things, who want to learn inside out. I think science is for them. There are lots of different things you can learn from science, but it's definitely not for me. I find it too much for my head.

In a similar manner, Hamiido, a woman in her 30s from the Somali group, simply laughed as she told me, "I don't know science". As well as embodied descriptions of having the wrong brain or head for science, some participants also commented on having the wrong bodies for science, namely that they were women.

As Mrs Mallick from the Asian group explained to me, most scientists were men, so how could she possibly be a scientist?

While the vast majority of participants had not studied science at school nor were they interested in science-related careers, they told stories of others who had tried and, like Mr Bhakta and Ibrahim in the next section, had struggled. For instance, Osmann from the Sierra Leonean group told me the story of his younger brother's thwarted science ambitions:

> My little brother, he's just 24, he graduated with a BSc in Environmental Studies, but he hasn't got a job to do, so he has to apply to a bank to work there, how can he? [. . .] So it's like, you see someone applying to university to do science, you say, "My friend, I did this thing. I haven't got a job. Why you do this thing? Go this way, there's no work there".

Stories like this circulated within the five groups and suggest that participants experienced science education and science careers as sites of struggle and, ultimately, worth avoiding in ways that mirror Bourdieu's (1998) work on symbolic violence. This symbolic violence operated even where participants were critical about how claims of cleverness and elitism worked in terms of science. For instance, Abdou from the Sierra Leonean group talked at length about his view that science teachers and other science professionals made science difficult on purpose,

> because they [science teachers and people in science-related jobs] come maybe from a rich background and they were fortunate to be successful, they're making it difficult so that nobody else can compete with them, so you know, it's a way to push people away. So they make science really, really difficult. [. . .] So that is the mentality we have, don't waste your time anyway, why do you want to get involved?

From Abdou's and Osmann's perspectives pursuing science was a waste of time, not least as a result of class discrimination. Although their resistance to elitism in science comes across in these two extracts, as Valerie Walkerdine's (1990) scholarship reminds us, not all resistance is revolutionary. To follow Bourdieu and Passeron's (1990) argument about symbolic violence, participants can be understood to have removed themselves from the system in anticipation of their rejection. However, their rejection of science education and careers from this perspective can be seen to reproduce existing classed and racialised patterns of participation within the scientific community.

Configuring science as the realm of the "super clever" also gave participants a way to resist their exclusion from science. Notably, people who *were* clever enough to participate in science were seen as distinct from normal people, as too clever and, as a result, not necessarily to be admired. Thus scientists were constructed as 'Other', as strange people making what participants' saw as dubious

decisions about animal testing or cloning. Hawa and Lucille from the Sierra Leonean group described this in terms of cultural beliefs in Sierra Leone, explain-
ing that being a "genius" was not necessarily positive, but somewhat alarming.
Indeed, participants in all five groups referred to scientists as "strange" and
"immoral" in ways that resonated with Hawa and Lucille's explanation. Idyl, for
instance, from the Somali group distinguished between what she saw as "normal"
people on one hand and scientists on the other. While these descriptions clearly
follow the pattern of positioning science as something special, there was an element of mockery and rejection of science in these accounts.

Participants' descriptions of science in general were overall somewhat of a
mixed bag and, while clearly patterned by their exclusion from science, not necessarily always negative. A pattern can be seen however in how science in general
was repeatedly configured as special and, as a result, exclusive (Michael, 1992).
Given that such views were expressed across five different community groups,
you can start to see how people might, over time, become disposed against science: science is overwhelmingly omnipresent, its wonders are too complicated to
understand and those who are involved in science are not like us and somewhat
dubious (Bourdieu & Wacquant, 1992).

Science in particular

Descriptions of science in particular worked to reconfigure science as something
participants could do or be actively involved with. Where science was relevant
to participants' lives they were passionate, interested, knowledgeable and highly
skilled. Amongst the Asian group for instance, assistive reproductive technologies were an example of science in particular about which the group were very
interested since one participant was undergoing fertility treatment. Similarly, at
the same time as describing her aversion to science in general and her hatred of
science while she had been at school, Flor from the Latin American group, talked
enthusiastically about the science involved in hairdressing, proudly stating, "yeah,
we do hair science":

> Like basically say, we look at a piece of hair, like one strand of hair, and then
> we go deep into what it is, like blah blah blah, chemical reactions and yeah,
> like to do with cuticles opening and closing and all that.

For Michael (1992, p. 313) "science-in-particular" was a device that rooted science in practical, identifiable efforts, reducing science to the status of the normal,
rather than the special. As the previous extract suggests, Flor was an accom-
plished hairdresser, highly skilled at dying hair and very knowledgeable about the
chemical processes involved. In talking about "hair science" Flor reconfigures science as something she can do, that she enjoys and that she has some expertise in.
Thus, as scholars across science education and communication have shown, and
continue to show, relevance and context matter in terms of how people perceive

science (Aguilar & Krasny, 2011; Barton et al., 2013; Marres, 2005; Roth & Tobin, 2007; Thompson, 2014).

Interestingly, for all that most participants talked about not pursing science careers, science hobbies or other science-related activities in general, thinking about science in particular provides a useful lens to reflect on how they were more knowledgeable and more practiced with science than they at first described themselves. We can think about Flor's description of "hair science" in terms of the asset based, funds of knowledge approach discussed in Chapter Three (Barton & Tan, 2009; Gonzalez, Moll, & Amanti, 2013). Flor's interest in and practice of science within hairdressing can be seen as a set of competences (knowledges and skills) in relation to an area of science in particular (Gonzalez, Moll, & Amanti, 2013; Michael, 1992). It is notable however, that Flor simultaneously highlighted her scientific skills and expertise whilst being somewhat dismissive of them ("blah blah blah"). Thus, configuring science as normal, mundane and something that participants could pursue was not wholly straightforward.

Participants accounts of science in particular suggest we need to be careful about how we contextualise their experiences. For instance, Hawa and Lucille from the Sierra Leonean group both worked as nurses and were clear about how nursing was, from their perspective, a scientific job. In contrast, Kirin from the Asian group who was a retired midwife, firmly rejected the idea that midwifery was a scientific job, insisting instead it was a domestic profession. Similarly, while some participants from the Asian group were, for example, adamant that cooking was not science, their extensive knowledge of different fish and fish habitats became very clear during our visit to an aquarium at a science museum we visited together. The fish were discussed in the context of how to find, prepare and cook fish for certain meals however, and this was *not*, as these participants later clarified, seen by them as scientific knowledge.

Contradictions in what participants saw as science suggest care is needed in how we understand the role of science in particular in people's lives, especially in terms of which kinds of practices are labelled as science, and by whom. Mrs Mallick from the Asian group was a prolific gardener, vastly knowledgeable about plants and botany. Ignacio from the Latin American group knew everything about Colombia's flora and fauna. The Sierra Leonean elders drew on a wealth of shared knowledge about health and well-being. I could continue to list examples. My point here is simply that participants were steeped in various forms of knowledge. Knowledges which are rarely understood as part of or even related to the scientific cannon. Knowledges which they themselves were sometimes eager to differentiate from science. Given participants' dispositions against science, framing their own knowledges and practices as science may have simply felt counterintuitive. If science was science because it was not for them but, in Mirza's words was "too much for my head" then perhaps, as Michael (1992) has argued, things they *could* do and *were* knowledgeable about simply couldn't be science?

Research about science and society relationships has long since argued that attitudes towards science are mediated by the perceived relevance of specific science topics (McKechnie, 1996; Michael, 2009; Solomon, 1993). But thinking about where science became mundane or normal for participants' needs to go further than trying to make science more relevant. It represents a challenge to how we understand whose knowledges and practices are and are not branded as science, along with implications for how science could be reconfigured.

Research about funds of knowledge and science capital all too clearly shows how powerful it can be to describe particular knowledges, skills or practices as science (Archer, Dawson, DeWitt, Seakins, & Wong, 2015; Barton & Tan, 2009). In terms of being able to use, build or exchange forms of capital, it is important in specific fields to be able to call something science (Archer et al., 2015; Bourdieu, 1984; Bourdieu & Passeron, 1990). At the same time however, making science less special seemed, for participants, to make it more recognisable and, as a result, more think-able and more do-able. Thus, reconfiguring science through the concept of science in particular, while paying attention to racialised, classed and gendered practices in whose knowledge counts, may offer different affordances for equity and inclusion in everyday science learning.

Being disposed against science

As I have shown in this section, participants' attitudes towards science were clearly more complex and nuanced than a simple dislike for science. Nonetheless, participants shared a disposition against science, a disposition born of a mixture of sometimes contradictory attitudes and experiences. Here it is helpful to think about attitudes to science as part of participants' habitus or tastes, or what Bourdieu described as "anticipations, a sort of practical induction based on previous experience" (1998, p. 80). From this perspective, Michael's (1992) concepts of science in general and in particular provide useful tools for understanding how participants' dispositions against science were able to incorporate multiple, shifting and contradictory perspectives about science, while retaining a generalised, shared habitus wherein science was seen as something for other people, and not for them. Of course, neither of these conceptual tools provide much space to consider how 'race'/ethnicity, class and gender might be implicated in people's attitudes to science.

Participants had many different experiences of science across their lives. But if we think about attitudes to science in terms of habitus and the space of lifestyles, a disposition against science could be shared across groups whose experiences are similar, as the data discussed here suggest. Despite specific, more positive experiences therefore, a disposition against science can develop against a backdrop of racialised, classed and gendered elitism.

Powerful messages about who science is and is not for disciplined how science was understood by participants, such that an orientation away from science became enduring and resilient (Bourdieu, 1990a, 1990b, 1998). A shared disposition against science does not grow from nowhere. It has socio-political historic roots. We ought to be careful therefore about seeing negative (or positive) attitudes towards science as just individualised character traits or opinions. Instead, as the stories in the next section make clear, we should think about the broader roles played by structural inequalities in terms of who in our societies learns that science is 'for them' and who learns that it is not.

Three stories about science

In this section I discuss in more detail the lives of the three participants who told me they liked science both in particular and in general. Their stories are based on narrative, life-history interviews we carried out together. I discuss these three stories here because, as in the previous section, they present a very different perspective on how exclusion from science operates than those based on deficit approaches to understanding exclusion. These three participants identified with science in general, really enjoyed aspects of science in particular and tried to pursue science in their own ways. Their positive attitudes towards science did not however, straightforwardly pave their way towards science careers or unlock the door to the various fields involved in everyday science learning.

In talking, as they did, about their lives in relation to science, participants' stories and silences help to highlight how structural inequalities, not least issues of colonialism, racism, class discrimination and sexism, affected their lives and their habitus in relation to science. It is certainly possible to examine exclusion from and non-participation in everyday science learning without asking broader questions about people's backgrounds or society, but to do so would be to miss significant pieces of this puzzle. Structural inequalities affect who can participate in science education and careers, as well as everyday science learning. As the extract from a conversation with Abdou from the Sierra Leonean group quoted in the previous section shows, this was common knowledge, though not necessarily talked about with me. Discussing people's life stories with a focus on their experiences of science highlighted again and again how paths towards science are paved very differently depending on who you are. And for some people, finding a path into science, in any field, is all but impossible.

These three stories come from participants from three different community groups, at different ages and with different ambitions. Despite their differences, the stories show how structural inequalities, not least colonialism and racism, continue to shape people's lives in relation to science and, in terms of the last story in particular, aspects of everyday science learning. As a result, these stories provide a backdrop against which participants' dispositions against science, as discussed in the previous section, can be understood.

Mr Bhakta's story: education, displacement and resilience

Mr Bhakta, a man in his 70s from the Asian group, traced his relationships to science through his life story in conversation with me one day, moving both of us to tears more than once. Mr Bhakta grew up in Uganda, the older of two sons, both of whom wanted to study medicine. His father had moved to East Africa from India as an indentured labourer, to build railways for the British. After the railways were built his father moved to Uganda, worked in a sugar factory on a sugar plantation and in 1939, after 10 years, set up his own business. Mr Bhakta and his two brothers moved between their parents and their extended family in India several times because, as he explained it, his father believed his sons would receive a better education in India. The role of education as a strategic tool, leveraged through transnational networks, can be seen here in Mr Bhakta's stories of his early life, and is a pattern he went on to repeat at university, and later with his own children and with his brother's children.

In his mid-20s Mr Bhakta returned to Mumbai to study medicine. As he put it "my ambition was to be a doctor, and if you become a doctor, you are always respected by the people. [. . .] It's a respected life when you are a doctor or a chemist". Mr Bhakta was interested in science in general as well as science in particular, and the high status of medical careers attracted him to medicine. Sadly, Mr Bhakta failed his exams at the end of the first year of his medical degree. Although he ultimately switched subjects and finished with a BSc in Chemistry and Biology, Mr Bhakta's regret and frustration about not becoming a doctor still coloured his story about his university experiences.

As Mr Bhakta explained it there were two routes into medicine in India when he was studying. The first route involved being very clever. You needed to achieve 90% or more in your exams and the competition was fierce. The second route involved being very rich. You needed to give a financial donation to a university and then it did not matter how clever you were. Mr Bhakta told me about other students he had known with worse exam results than his, who were protected through their family wealth and went on to become doctors. He was in no doubt about how this system operated: you could buy your medical degree. Except, of course, that he could not afford to. His younger brother had a similar experience trying to study engineering in Mumbai a few years later and also switched to a BSc in Chemistry and Biology.

After graduating with his BSc, Mr Bhakta returned to Uganda to teach biology and maths at a secondary school. He got married and his first daughter was born just before Idi Amin came to power. Idi Amin's violent persecution of ethnic minorities in Uganda led to the expulsion of everyone who was not a Ugandan citizen. With their British passports, the result of the struggle for independence from the British Empire, Mr Bhakta and his family left Uganda with only £50. They arrived in Heathrow in the snow in November 1972 and lived in a refugee camp set up at an army base. Mr Bhakta's family rebuilt their lives, with members of the wider family moving, one by one, to London. Mr Bhakta found he was unable to teach science or maths in the UK because his qualifications were "foreign", so he ran a small newsagents shop until he retired aged 65. He was successful enough with the business that he was able to support not only his immediate family, but the family of his younger brother, who had died aged 32 leaving a wife and two small children in India. Mr Bhakta brought them to

London. Mr Bhakta, despite his own set-backs, still saw science careers as high status occupations and maintained science career aspirations for his family. He saw his nephew's successful career as a pharmacist in London as a huge cause for celebration. At last, someone in his family had been able to pursue his dreams of a science-related career.

Let's think about this story for a moment. Mr Bhakta's family move from India, across Africa, to Uganda, between Uganda and India and, ultimately, to the UK. This is a story marked by the complex and damaging legacy of British colonialism and the socio-political ramifications of that history in the 20th and 21st centuries. As Mr Bhakta and his brother tried to pursue their medical ambitions, theirs became a story marked by people who, in relation to science in general and in particular, were framed in terms of deficits, as not rich or clever enough. As Mr Bhakta described his science education experiences, both as a student and later a teacher, this is a story marked by political deportation, poverty and rebuilding an alternative career to support a family, when "foreign" qualifications meant he could no longer teach science and maths. Thus issues of racism, colonialism and socio-economic status are interwoven in Mr Bhakta's life story in ways that meant he could not pursue his two different chosen science-related careers, despite his interests in science, his ambitions and his evident capacity for hard work. Indeed, Mr Bhakta's story is useful because we can trace how structural inequalities affected his life, not least in terms of his capacity to identify with, enjoy and work in science.

Ibrahim's story: interests and opportunities

Ibrahim, a man in his 40s from the Sierra Leonean group, was almost 30 years younger than Mr Bhakta, but their stories are not as dissimilar as you might first think. Ibrahim grew up in Sierra Leone and studied science at university in Freetown, graduating with an MSc in Biology. After his MSc he worked in various countries in East Africa, carrying out environmental research projects and eventually settled in Freetown. Ibrahim was proud of his science education. He and his friends joked that his MSc in Biology was highly unusual and that few of their acquaintances could claim similar qualifications. For Ibrahim, "science is in everything, science is wonderful". Like Mr Bhakta he was interested in science in general and passionate about science in particular (Michael, 1992). In his own words, Ibrahim was someone who enjoyed "science and nature".

Freetown, the capital of Sierra Leone, saw a significant proportion of the violence during the Sierra Leonean civil war of the 1990s. Like many of his friends and family, Ibrahim moved from Freetown to the UK to escape the fighting. Later, like Mr Bhakta, he moved regularly backwards and forwards between the two countries he called home. Many of the Sierra Leonean participants commented on the enduring effects of the civil war in Sierra Leone, the damage to the country's infrastructure, limited funding for health care, let alone education, and many, like Ibrahim, regularly returned frequently to help with what they called "rebuilding" their country.

In the UK despite his passion for science and describing himself as a scientist, Ibrahim, worked as a security guard for a chain of high-street shops. Abdou, his close friend, explained how Ibrahim ended up in this job:

ABDOU: He (Ibrahim) has got environmental science background but he cannot get a job within that field because it's not catered for in Sierra Leone and even here because he did not study here it would be very difficult for him to break through the ranks you know and he's living in a poorer community. He's got interest which is fine, but you know, for him to get the opportunities to pursue what he wants to do . . . [pause]
EMILY: At work?
ABDOU: And to work, it's not there, you know. It's like wasted resources, spending four years in university studying science, when you come back you have to go and do security job or something else that's not related, so that demotivates you or others.

Abdou told me his views about Ibrahim's current job after we had been talking about Ibrahim's passion for a variety of different areas of science. His views echo those discussed in the previous section, Abdou just could not see the point in studying science when it seemed impossible to get a science-related job and his sense of frustration on behalf of his friend was palpable.

Ibrahim had been the only participant amongst the Sierra Leonean group who described liking science in general (Michael, 1992). Even Hawa and Lucille, both nurses who considered their jobs to be scientific jobs and were extremely knowledgeable about medicine, talked about not liking science in general. Ibrahim, unlike everyone else in his community group, told me that although he did not know where the Science Museum was in London, he would visit it if he could, something his friends found laughable. But Abdou's cynicism about science-related opportunities was not uncalled for. As he pointed out, like Mr Bhakta, Ibrahim's qualifications were devalued because they came from Sierra Leone. He found he was unable to pursue the work he had been trained for. And despite living in central London on and off since the early 1990s, Ibrahim did not know where the Science Museum was. Thus, it is hard not to agree with Abdou's comment that the opportunities to pursue science are not there for people in his community.

Like Mr Bhakta, Ibrahim's story is marked by socio-political histories of colonialism, warfare, globalisation, hardship and personal resilience. That Mr Bhakta and Ibrahim both remained passionate about science in general and in particular *and* continued to identify with science was testament to their tenacity. It is important therefore to think about the pattern Mr Bhakta and Ibrahim's stories exemplified. Their individual experiences fit within research on the "brain drain" effects of migration (Ahmad, 2004; Escadón, 2017). Their stories also fit within research on post-colonialism, transnational migration and super-cities that has shown how practices rooted in colonialism and racism shape (and limit) the opportunities of people from racialised backgrounds who move to places like London (Gilroy, 2002; Modood, 2010; Sassen, 2001; Vertovec, 2007). In both the 1970s and the 1990s, devalued "foreign" qualifications did not support Mr Bhakta or Ibrahim to pursue employment in the scientific work they had been trained to do. Both participants came from countries where British colonialism had deep roots, tied to economic patterns that favour the Global North at the expense of the Global South. As their stories show, structural inequalities had enduring and damaging effects on the lives of both Mr Bhakta and Ibrahim, effects that limited or wholly undermined their science-related ambitions. As Ibrahim's friend Abdou concluded, "he's got an interest, which is fine, but it's not enough".

Fatima's story: experiences of everyday science learning

I turn now to Fatima's story because her story moves away from science in formal education and careers to focus more on science in hobbies and everyday science learning activities. Thus, while Ibrahim insisted he would visit the Science Museum in London if he only knew about it and could find it, Fatima argued quite the opposite. Despite her interest in science and her pursuit of various other forms of everyday science learning activities, Fatima, in her words "hated" science museums.

When I met Fatima she was in her mid-20s and been involved with the Somali community group for several years. Like Mr Bhakta and Ibrahim, Fatima's presentation of herself was at least partially built around her science interests and skills. In other words, science in general and in particular featured in her "practiced identity" (Holland, Skinner, Lachiotte Jr., & Cain, 2001, p. 271; Michael, 1992). Unlike Mr Bhakta and Ibrahim, Fatima had grown up in the UK, had gone through the British school system in London, was ambivalent about science at school and had no ambitions to pursue science as a career. She did however find science interesting and pursued it in several ways.

Fatima described herself as the "odd one out of the family; I'm a weirdo" because she was interested in science and preferred staying in reading books to going out. It turned out that the kinds of books Fatima read were also unusual from her perspective compared to what her friends and family enjoyed reading. Fatima read science books, specifically books on her twin passions for science in particular; engineering and biology. Against the context provided earlier in this chapter it is easy to see how being interested in science in general or in particular seemed unusual to Fatima and that her hobbies, in turn, seemed unusual to her friends and family.

Not only was Fatima into science and reading books about it, she pursued her interest in science through other forms of everyday science learning. She purposefully sought out ways to develop her knowledge of science through specific practices that echo participants' use of the more accessible fields of everyday science learning discussed in Chapter Three. For instance, she talked about going online to research aspects of engineering and biology and being known amongst family and friends as good at finding useful scientific information when it was needed. While she watched science on television, unlike other participants discussed in Chapter Three, she went out of her way to watch programmes with a lot of science content, particularly enjoying science documentaries and nature programmes. The fields of everyday science learning she was involved with however, as discussed in Chapter Three, were the more popular, low-brow or mundane practices, ones that she could pursue in her own time, in her own home. In contrast, dominant, high-brow, elite practices of everyday science learning, particularly visiting museums, were not something she enjoyed; rather they were something she sought to avoid.

In one of our first meetings Fatima bluntly told me: "I hate museums". In a later interview she continued: "I'm very upset with the museums, so I'm not going. . . . I just did it because I had to do it at school". Fatima described school visits to museums in negative terms as "a sort of detention" and "punishment". Unlike most of the people involved in the research carried out for this book, Fatima was able to draw on her previous experiences of informal science learning at length because she had visited several museums and similar institutions in London, including the Natural History Museum, the British Museum, the Science Museum, London Zoo and Vauxhall City Farm. But these visits did not mean Fatima liked science museums or similar institutions. On the contrary, she told me she thought their outreach practices failed to meet those they should (all of the public) and were simply not up to standard. She said she had never seen an advert for a science museum in her neighbourhood, nor leaflets, signs, or information in community newspapers, websites or on radio stations and that she felt her community had been left to one side as a result.

Fatima's underlying assumption was that visiting a science museum or anything like it would be unusual for her, her friends, family and broader community. As she put it, "I don't know anyone that's decided one day 'oh, let's go to the museum'". Her experiences generated a story about science museums and similar institutions as, from her perspective, poor-quality, off-putting and irrelevant. This was a story reinforced and reproduced socially amongst her community and friends, into a world where science museums were inaccessible, unpleasant and removed from day-to-day life.

Despite her personal interest in science (in particular and in general) and her enjoyment of other science-related hobbies and everyday science learning practices, visiting a science museum was not a choice Fatima expected to make, nor did she expect her friends or family to do so, as the extract that follows shows:

FATIMA: if you don't know anything about the museum and it's not part of your social outlook then you don't know what's happening in the museum.
EMILY: Yeah, and you'd never look it up?
FATIMA: You'd never look it up, you wouldn't have no need to, because it's not something you do.

This extract speaks to a deeply ingrained sense of not belonging in science museums. Fatima's story highlights the complexity of people's lives and a concrete context for thinking about how a person might navigate the various fields involved in everyday science learning. Engaging with science from a marginalised social position creates cross-cultural differences that require considerable negotiation and can produce multiple, heterogeneous identities (Aikenhead, 2002; Roth, 2008). Fatima worked hard to find ways to describe what, at times, felt like contradictory dispositions towards everyday science learning, dispositions that make more sense when understood in context of which fields of everyday science learning were more or less accessible.

Fatima's story provides a particularly useful example of how rejecting certain forms of everyday science learning can go hand in hand with being excluded from those same practices. Fatima's response to feeling excluded from science museums and zoos was to reject them (a pattern discussed more in Chapter Six). Though she felt her interest in science was unusual, as was her pursuit of science books and documentaries, Fatima's experiences of and attitudes towards science museums were broadly in line with those of the other participants. She shared the disposition that oriented participants away from dominant forms of everyday science learning as practices that were, as discussed in Chapter Three, by and large unheard of, unusual, unhelpful at best and, as explored in the next two chapters, damaging at worst (Bourdieu, 1998; Bourdieu & Wacquant, 1992).

As in the stories of Mr Bhakta and Ibrahim, symbolic violence can be identified in Fatima's story. Symbolic violence can be understood as "based on 'collective expectations' or socially inculcated beliefs" (Bourdieu, 1998, p. 103). It sneaks into well-meant intentions, in doing what you have to do, usually do and expect to do. As Fatima put it, "why do something you don't, it's not part of you". Thus, in Fatima's story, her rejection of science museums worked hand in hand with the structural inequalities that marginalised her in such spaces, a disturbingly good example of symbolic violence. Fatima's rejection of science museums, in combination with their inaccessibility, created a resilient system of exclusion that would be hard for Fatima to change, even if she wanted to.

Fatima's attitudes to science (both in general and in particular) and her differentiated involvement in everyday science learning usefully challenge the deficit approach to understanding exclusion from everyday science learning. Fatima did not like science because of her visits to science museums, but rather, in spite of them. From this perspective, liking science in general and in particular paved her way towards the more accessible everyday science learning practices that could be used or enjoyed from home. It was not, however, enough to make her feel at home in a science museum, for reasons discussed in more detail in the next two chapters. Framing Fatima's exclusion from everyday science learning as the result of personal opinions and knowledge deficits simply does not work.

Learning that science is not for you

It is important to think carefully about the experiences these three stories and the perspectives of participants towards science in general and in particular exemplify.

In many ways each story about science at school, at university, as a career and as a hobby seems unique. Each story is the product of a very specific time and place, a story that affects specific people in specific ways. But what if we compare the stories to one another and the experiences of participants more broadly? They share themes about how structural inequalities shape the science-related opportunities that people can access. They speak to the naiveté of assuming that an interest in science and identifying with science (in general or in particular) will be enough to support the science-related ambitions of people from minoritised groups.

The point I am trying to make here, is that for people across five different community groups to share a sense that science was not for them, from school experiences in their childhood onwards, suggests that their orientation away from science in general is not simply the result of an individual experience or preference. Remember, participants' shared characteristics were of migration, of racialised and economic disadvantage in the UK. Racialised groups have a long history of struggle in the UK, one marked by colonialism, racism, class discrimination and sexism (Gilroy, 2002; Puwar, 2004; Virdee, 2017). Though members of each community group had their own stories to tell, in tracing experiences of science through their lives, in and out of education systems and beyond, it is possible to see how shared dispositions against science develop.

That participants felt quite differently about science 'in particular' compared to science 'in general' suggests a need to reconsider what kinds of knowledges are reified within and beyond everyday science learning practices. What we usually describe as the traditional canon of scientific knowledge was made by and for wealthy white men. Epistemic practices about what kinds of knowledge and practice count as science disciplined participants' discourses. This finding does not simply mean that everyday science learning practices ought to reframe science around the familiar features of people's everyday lives. This ought to already be a standard feature of good practice (see for example, Aikenhead, 2002; Barton et al., 2013; King & Nomikou, 2018). I argue instead that we also need to consider what counts as *science* in a more fundamental way. To open up the idea of science 'in particular' to make space for how racialised, classed and gendered the construction of science has been. Rethinking what counts as science may create new affordances for equity and inclusion in everyday science learning.

Participants' stories point towards systemic problems about how certain groups – or publics – are constructed through institutional and community practices in relation to science as unknowledgeable, excluded and Other. As a result, a broader, more nuanced account of exclusion from science and everyday science learning is required; one that goes beyond individual tastes or interests in science, to think instead about systematic practices of inaccessibility and exclusion.

Summary

This chapter discussed participants' attitudes towards and experiences of science both as themes across all five community groups and in the context of three specific participants' lives. Throughout this chapter I argued that deficit approaches to understanding exclusion in everyday science learning premised on negative attitudes towards science do not hold up when we look at participants' experiences in more detail. Indeed, in this chapter I have shown how an interest in science or being able to identify with science does not necessarily unlock the door to being able to participate in science, whether in formal education, as a career or via everyday science learning.

The data in this chapter suggest instead that structural inequalities limited and undermined participants' access to and pursuit of science-related activities in various ways. Drawing on Bourdieu's (1992) concept of habitus and Michael's (1992) work on discursive formulations of science, I have argued that we ought instead to understand participants' shared dispositions against science as structured by systematic inequalities that cut across their experiences of science in ways that are pernicious and enduring. The socio-political and historic roots of structural inequalities embedded across the different systems participants encountered in their lives, from education to employment, are not to be taken lightly. Assuming that individualised attitudes towards science are the locus of the problem of exclusion from everyday science learning is, from this perspective, absurd. Instead, as this chapter has shown, we need to reconsider participants' exclusion from everyday science in the context of accumulated, shared experiences that positioned science 'in general' as unthinkable for participants.

References

Agar, J. (2012). *Science in the 20th century and beyond*. Cambridge: John Wiley & Sons.

Aguilar, O. M., & Krasny, M. E. (2011). Using the communities of practice framework to examine an after-school environmental education program for Hispanic youth. *Environmental Education Research*, *17*(2), 217–233. doi:10.1080/13504 622.2010.531248

Ahmad, O. B. (2004). Brain drain: The flight of human capital. *Bulletin of the World Health Organization*, *82*(10), 797–798.

Aikenhead, G. (2002). Science communication with the public: A cross-cultural event. In W.-M. Roth & J. Désautels (Eds.), *Science education as/for sociopolitical action* (pp. 151–166). New York: Peter Lang Publishing.

Archer, L., Dawson, E., DeWitt, J., Seakins, A., & Wong, B. (2015). "Science capital": A conceptual, methodological, and empirical argument for extending bourdieusian notions of capital beyond the arts. *Journal of Research in Science Teaching*, *52*, 922–948. doi.org/10.1002/tea.21227

Archer, L., DeWitt, J., Osborne, J., Dillon, J., Willis, B., & Wong, B. (2010). "Doing" science versus "being" a scientist: Examining 10/11-year-old schoolchildren's constructions of science through the lens of identity. *Science Education*, *94*(4), 617–639. doi.org/10.1002/sce.20399

Archer, L., Nomikou, E., Mau, A., King, H., Godec, S., DeWitt, J., & Dawson, E. (2019). Can the subaltern "speak" science? An intersectional analysis of performances of "talking science through muscular intellect" by "subaltern" students in UK urban secondary science classrooms. *Cultural Studies of Science Education, Online First.*

Barton, A. C., Kang, H., Tan, E., O'Neill, T. B., Bautista-Guerra, J., & Brecklin, C. (2013). Crafting a future in science: Tracing middle school girls' identity work over time and space. *American Educational Research Journal, 50,* 37–75.

Barton, A. C., & Tan, E. (2009). Funds of knowledge and discourses and hybrid space. *Journal of Research in Science Teaching, 46*(1), 50–73. doi.org/10.1002/tea.20269

Bourdieu, P. (1984). *Distinction: A social critique of the judgement of taste* (R. Nice, Trans.). London: Routledge.

Bourdieu, P. (1990a). *In other words: Essays towards a reflexive sociology.* Stanford: Stanford University Press.

Bourdieu, P. (1990b). *The logic of practice* (R. Nice, Trans.). Stanford: Stanford University Press.

Bourdieu, P. (1998). *Practical reason.* Cambridge: Polity Press.

Bourdieu, P., & Passeron, J.-C. (1990). *Reproduction in education, society and culture* (R. Nice, Trans., 2nd ed.). London, Newbury Park, CA and New Delhi: Sage.

Bourdieu, P., & Wacquant, L. (1992). *An invitation to reflexive sociology.* Chicago: University of Chicago Press.

Carlone, H. B., Scott, C. M., & Lowder, C. (2014). Becoming (less) scientific: A longitudinal study of students' identity work from elementary to middle school science. *Journal of Research in Science Teaching, 51,* 836–869.

Chaffee, R., & Gupta, P. (2018). Accessing the elite figured world of science. *Cultural Studies of Science Education, 13*(3), 797–805. doi:10.1007/s11422-018-9858-0

Dawson, E., Archer, L., Seakins, A., DeWitt, J., Godec, S., King, H., Mau, A. & Nomikou, E. (2019). Selfies at a science museum: Exploring girls' identity performances in a science learning setting. *Gender and Education.*

Escadón, T. A. (2017). *Scientific migration and the brain drain in Mexico.* (PhD Monograph). London: University College London.

Gilroy, P. (2002). *There ain't no Black in the Union Jack* (2nd ed.). Abingdon: Routledge.

Gonzalez, N., Moll, L. C., & Amanti, C. (2013). *Funds of knowledge: Theorizing practices in households, communities, and classrooms.* Mahwah, NJ: Taylor & Francis.

Hammonds, E. (2009). *The nature of difference: Sciences of race in the United States from Jefferson to genomics.* Cambridge, MA: MIT Press.

Haraway, D. (1997). *Modest_Witness@Second_Millennium.FemaleMan©Meets_OncoMouse™: Feminism and technoscience.* New York: Routledge.

Hodari, A., Ong, M., Ko, L., & Smith, J. M. (2016, May/June). Enacting agency: The strategies of women of color in computing. *Computing in Science & Engineering,* pp. 58–68.

Holland, D., Skinner, D., Lachiotte Jr., W., & Cain, C. (2001). *Identity and agency in cultural worlds.* Cambridge, MA and London: Harvard University Press.

Hussénius, A. (2014). Science education for all, some or just a few? Feminist and gender perspectives on science education: A special issue. *Cultural Studies of Science Education, 9*(2), 255–262. doi:10.1007/s11422-013-9561-0

King, H., & Nomikou, E. (2018). Fostering critical teacher agency: The impact of a science capital pedagogical approach. *Pedagogy, Culture & Society, 26*(1), 87–103. doi:10.1080/14681366.2017.1353539

Knorr Cetina, K., & Mulkay, M. (1983). Introduction: Emerging principles in social studies of science. In K. Knorr Cetina & M. Mulkay (Eds.), *Science observed: Perspectives on the social study of science* (pp. 1–17). London, New Delhi and Beverly Hills: Sage.

Lock, S. J. (2011). Deficits and dialogues: Science communication and the public understanding of science in the UK. In D. J. Bennett & R. C. Jennings (Eds.), *Successful science communication: Telling it like it is* (pp. 17–30). Cambridge: Cambridge University Press.

Marres, N. (2005). *No issue, no public: Democratic deficits after the displacement of politics.* (Phd Monograph), University of Amsterdam, Amsterdam.

McKechnie, R. (1996). Insiders and outsiders: Identifying experts on home ground. In A. Irwin & B. Wynne (Eds.), *Misunderstanding science? The public reconstruction of science and technology* (pp. 126–151). Cambridge: Cambridge University Press.

McNeil, M. (2013). Between a rock and a hard place: The deficit model, the diffusion model and publics in STS. *Science as Culture, 22*(4), 589–608.

Medin, D. L., & Bang, M. (2014). *Who's asking? Native science, western science, and science education.* Cambridge, MA and London: MIT Press.

Michael, M. (1992). Lay discourses of science: Science-in-general, science-in-particular, and self. *Science, Technology & Human Values, 17*(3), 313–333. doi:10.1177/016224399201700303

Michael, M. (2006). *Technoscience and everyday life: The complex simplicities of the mundane.* Maidenhead and New York: Open University Press.

Michael, M. (2009). Publics performing publics: Of PiGs, PiPs and politics. *Public Understanding of Science, 18*(5), 617–631.

Modood, T. (2010). *Still not easy being British: Struggles for a multicultural citizenship.* Stoke on Trent: Trentham Books.

Ong, M., Wright, C., Espinosa, L. L., & Orfield, G. (2011). Inside the double bind: A synthesis of empirical research. *Harvard Educational Review, 81*(2), 172–390.

Osborne, J., Simon, S., & Collins, S. (2003). Attitudes towards science: A review of the literature and its implications. *International Journal of Science Education, 25*(9), 1049–1079. doi.org/10.1080/0950069032000032199

Puwar, N. (2004). Thinking about making a difference. *The British Journal of Politics and International Relations, 6*(1), 65–80. doi:10.1111/j.1467-856X.2004.00127.x

Ravetz, J. (2005). The post-normal safety of science. In M. Leach, I. Scoones, & B. Wynne (Eds.), *Science and citizens: Globalization and the challenge of engagement* (pp. 43–53). London: Zed Books.

Roth, W.-M. (2008). Bricolage, métissage, hybridity, heterogeneity, diaspora: Concepts for thinking science education in the 21st century. *Cultural Studies of Science Education, 3*(4), 891–916. doi.org/10.1007/s11422-008-9113-1

Roth, W.-M., & Tobin, K. (2007). Aporias of identity in science: An introduction. In W. M. Roth & K. Tobin (Eds.), *Science, learning, identity: Sociocultural and cultural-historical perspectives* (pp. 1–10). Rotterdam and Taipei: Sense Publishers.

The Royal Society. (1985). *The public understanding of science.* London: The Royal Society.

Sassen, S. (2001). *The global city: New York, London, Tokyo* (2nd ed.). Princeton, NJ and Oxford: Princeton University Press.

Solomon, J. (1993). Reception and rejection of science knowledge: Choice, style and home culture. *Public Understanding of Science, 2*(2), 111–121. doi:10.1088/0963-6625/2/2/002

TallBear, K. (2013). Genomic articulations of indigeneity. *Social Studies of Science*, *43*(4), 509–533. doi:10.1177/0306312713483893

Thompson, J. (2014). Engaging girls' sociohistorical identities in science. *Journal of the Learning Sciences*, *23*, 392–446. doi.org/10.1080/10508406.2014.888351

Ulriksen, L., Madsen, L. M., & Holmegaard, H. T. (2010). What do we know about explanations for drop out/opt out among young people from STM higher education programmes? *Studies in Science Education*, *46*(2), 209–244.

Vertovec, S. (2007). Super-diversity and its implications. *Ethnic and Racial Studies*, *30*(6), 1024–1054.

Virdee, S. (2017). The second sight of racialised outsiders in the imperialist core. *Third World Quarterly*, *38*(11), 2396–2410. doi:10.1080/01436597.2017.1328274

Walkerdine, V. (1990). *Schoolgirl fictions*. London: Verso.

Ziman, J. M. (2002). *Real science: What it is, and what it means*. Cambridge: Cambridge University Press.

Chapter 5

Feeling excluded

It is quite confusing I think. I like the idea of liking the [science] things, but actually to do them is something else. So I like the sense of going to the zoo, purely just to see the animals, but I wouldn't. [. . .] So you said I like how things are built and how things are, so seeing an animal in a cage, it gets my brain thinking, oh the journey that the animal had, like to get a lion from the jungle, to get it to the zoo, the cage, and see how the lion actually adapts, 'cos it's a different scenario for the lion, but [. . .], in terms of the museums and stuff like that, it's not my cup of tea at all (laughs).

What does it mean when everyday science learning practices such as visiting zoos and museums are, as Fatima from the Somali group put it in the previous interview extract, not your cup of tea? As she herself suggests, it can seem quite "confusing", especially when, like Fatima, you like science. It is however, perhaps not so confusing when you remember that, as discussed more in this and the next chapter, Fatima and the other participants rarely saw themselves represented in everyday science learning, and when they did, it was typically in ethnographic exhibits or 'outreach' programmes, where they were objectified.

As post-colonial theorists have asked, what does it mean for people from racialised groups to grapple with and respond to issues of representation, cultural imperialism and colonialism (Bhabha, 1994; Gilroy, 2002; Spivak, 1988, 1999)? Their perspective is likely different to that of dominant groups. As Fatima put it, empathising with the objectified animals on display in zoos, "it's a different scenario for the lion". Understanding exclusion and non-participation from participants' perspectives provides an alternative view on equity, exclusion and everyday science learning, one that, as discussed in Chapters One, Two and Three, rarely makes its way into the research literature.

In this chapter I explore how participants framed and understood their own non-participation and exclusion from everyday science learning. I argue that exclusion and non-participation were visceral, embodied experiences for participants. In other words, they *felt* excluded. To understand their embodied dispositions I build on ideas from social justice (particularly cultural imperialism and powerlessness) and Bourdieu (particularly habitus and symbolic violence). I begin by discussing how non-participation and exclusion might be understood as embodied. In the second section I discuss participants' exclusion from everyday science learning in terms of oppression (cultural imperialism and powerlessness). In the third section I discuss how participants imagined the kinds of publics involved in everyday science learning and in the fourth section I focus in on how participants imagined science museums (and similar spaces) and their publics in relation to their own selves.

Research on everyday science learning has been constrained by an inward focus on those who do participate at the expense of those who do not. As a result, comparatively few studies have explored the attitudes and experiences of people who are not willing and/or able to participate in everyday science learning. Data on the views of excluded or non-participating groups about *why* they are not more involved in everyday science learning is particularly scarce. We know an awful lot about the motivations of science centre or zoo visitors, how school students learn science in museums or how people attempt to influence science policy consultations (de Saille, 2014; DeWitt & Osborne, 2010; Falk & Gillespie, 2009). But, given the issues discussed in this book so far, I think we ought to be cynical about how transferable the findings of some of these studies are to populations beyond those who already access everyday science learning and who already find their interests represented, respected and valued.

Structural inequalities and exclusion from everyday science learning

How can we understand the specifics of what it feels like to be excluded from everyday science learning? In this section I explore participants' responses to their position in everyday science learning and argue that participants' exclusion can be understood in terms of embodied exclusion, cultural imperialism and powerlessness (Fraser, 2003; Spivak, 1988; Young, 1990a). These three features of oppression highlight how structural inequalities, particularly those rooted in racism, reinforced exclusion from everyday science learning across the intersecting subjectivities of participants' lives.

Embodied exclusion

Exclusion was understood in embodied terms by participants. That is, participants located their exclusion and non-participation in the relationships between their racialised, classed and gendered bodies and the kinds of bodies they imagined everyday science learning catered for. It was something they felt and something they lived. Given the patterns discussed in Chapters Three and Four, it is perhaps not surprising that participants steered clear of dominant everyday science learning activities or, that with few exceptions, they rarely pursued the more accessible, popular forms of everyday science learning either.

For participants, science and everyday science learning were historically, politically, socially and culturally constructed as spaces for people who were, as Mirza from the Asian group concluded, "not like us". As Bourdieu and Passeron (1990) have argued, the most effective form of domination is that which "comes from exclusion, which perhaps has the most symbolic force when it assumes the guise of self-exclusion" (1990, pp. 41–42). In this sense, combining ideas about habitus and structural inequalities, participants' involvement in everyday science learning was framed as hard to imagine, unthinkable, unlikely, and, given its oppressive features, unwanted. Thus, while it is vital to unfold how power and exclusion operate in everyday science learning, so too must room be left for people to genuinely reject participation in everyday science learning, even if *at the same time* practices are oppressive such that they would be excluded anyway.

To understand how exclusion and/or non-participation work and, indeed, how they are interconnected and, in certain ways, inseparable, we have to think beyond numbers and turn instead to people's expectations and experiences (Puwar, 2004; Saha, 2018). Institutions and their practices, whether in schools, universities, workplaces or museums, are renowned mechanisms of social reproduction. As such, we need to pay attention to questions of belonging, who feels welcome and unwelcome in everyday science learning practices and institutions, what Ahmed describes as "how some more than others will be at home in institutions that assume certain bodies as their norm" (2012, p. 3). Thus, whether in novels, universities, on television or in museums we must think critically about representation, not least in terms of how representation is racialised, classed and gendered, as well as how these intersect (Hall, 2013; McIntyre, 1997; Spivak, 1999). How appealing is it to visit a science centre, festival or museum where no one else looks like you, except sometimes in an ethnography gallery? How much more work is involved in feeling comfortable with a practice when you are not welcomed or respected? How exhausting is the emotional work of watching a television programme about science that offensively misrepresents your culture? How appealing might everyday science learning seem under these conditions?

Spaces are at once both real and imagined. As Young (1990b) has argued, for those at the sharp end of structural inequalities, both the real and the imagined discipline the body in particular ways. Thus, institutions and their practices can be comfortable for those privileged enough to never need to notice how well their experiences in the world align with their needs, interests and dreams (McIntyre, 1997). In contrast, as numerous scholars have discussed, people from racialised groups, working-class and socio-economically disadvantaged backgrounds, women and girls, as well as those who inhabit multiple aspects of these subjectivities, must sometimes tread more carefully through these spaces, and at greater cost[1] (see for example Crenshaw, 1991; Fanon, 2008/1967; Lorde, 1984; Massey, 1994; Moraga & Anzaldúa, 2015/1981; Puwar, 2004; Young, 1990b).

Such costs involve not only the more obvious disadvantages that arise from structural inequalities – the opportunities lost, the higher price paid for mistakes – but an embodied, emotional labour, the extra work required to 'fit' and the discomfort of not 'fitting' (Ahmed, 2017; Hochschild, 1979; Holland, Skinner, Lachiotte Jr., & Cain, 2001; Lorde, 1984; Scott, 2009). And to be clear, the additional emotional costs and work required to put themselves into potentially disadvantageous everyday science learning situations were not appealing to participants.

Cultural imperialism

Cultural imperialism – when the culture, views and practices of the socially dominant appear universal at the expense of the marginalised – was particularly salient for participants in terms of 'race'/ethnicity (Young, 1988, 1990a). In the UK, socially dominant cultural practices – including representation in terms of language, stories, knowledge and images – can be understood as racialised. Specifically, the knowledges, practices and the representation of people from white ethnic, middle- and upper-class backgrounds are valorised, and those practices reserved for men remain at a premium (Beck, 1995; Belfiore, 2018; Hall, 2013; Massey, 1994; McIntosh, 1989; Miles & Gibson, 2016; Puwar, 2004). This was no secret to participants.

Participants saw everyday science learning practices as Eurocentric and reproducing racist stereotypes. Participants in the Somali and Sierra Leonean groups described for instance, how they resented the perception of Africa as burdened by disease and "saved" by the West in stories about medicine. White saviour narratives are not uncommon across culture, policy and education, but they are profoundly misrepresentative, disempowering and racist (Kendall, 2015; Puwar, 2004).

Issues of 'race'/ethnicity and cultural imperialism also intersected with gender. For example, three participants from the Sierra Leonean and Asian groups – all nurses – disputed the celebration of Florence Nightingale against the comparative invisibility of Mary Seacole (who they saw as a positive example of Black womanhood and nursing) on television and in science museums. For Hawa and Lucille from the Sierra Leonean group, how Black nurses and Black people were represented were part and parcel of a long list of racist forms of representation embedded as normal practices in everyday science learning:

HAWA: Because in reality, a lot of Africans have done a lot of things that are good in the world. But most of the time when people are talking about history, when you think about science in the museums, they are forgotten. Maybe the only good thing they put about Black people is in nursing, Florence Nightingale and Mary Seacole, or the slavery.

LUCILLE: Even Mary Seacole, it's a major example, Florence Nightingale, they portray her so much in the world.

HAWA: And Mary Seacole is forgotten [. . .] because the things they will tell you about are slavery [. . .] like the media, when they go to Africa, they take photos in the worst areas.

LUCILLE: And that negativity of portrayal, like, wait a minute, if you are there and you have friends there, why can't they show us in a different sort of [light].

Issues of gender and 'race'/ethnicity intertwine here. Hawa and Lucille described what they saw as a whitewashed, disempowering history in terms of slavery, science, nursing and Black women. Women who, as members of racialised groups, have struggled historically and today to have their lives recognised and respected (see for example Bassel & Emejulu, 2017; Hill Collins & Bilge, 2016; hooks, 1981). From this perspective, the "negativity of portrayal" that Hawa and Lucille refer to reflects dominant racialised and gendered discourses that situated them as both deficient and Other (Bhopal, 2018; Young, 1990b).

Evident in participants' accounts of cultural imperialism, from a Bourdieusian perspective, is that participants did not lack cultural capital. Rather, their cultural capital – the stories, practices and knowledge they valued – was not reflected in the everyday science learning landscape as they saw it. Building on research about white privilege, whiteness as a form of property and whiteness as a valued form of cultural capital, participants experiences suggest everyday science learning is racialised such that cultures, knowledges and practices associated with whiteness are valorised (Bell, 1992; Bhopal, 2018; Hage, 1998; Harris, 1993; McIntosh, 1989). As a result, people from racialised groups face an extremely uneven playing field within and beyond everyday science learning.

(Mis)representation mattered in participants' accounts of exclusion from everyday science learning. For example, Connie from the Afro-Caribbean group felt science museums pigeonholed Black people via narrow, racist portrayals. In the same vein, she argued everyday science learning activities tailored to her community during Black history month were tokenistic, angrily stating, "we're not invited the rest of the year!" Maria from the Latin American group told me a similar story about how frustrated she had been when her community group were asked by a prestigious London museum to be part of their Day of the Dead celebrations. No language provision was made for her friends who were less fluent in English and community artists (dancers and musicians) were expected to perform for free, without even their food or travel expenses being covered.

Connie's and Maria's stories of cultural imperialism can be understood as what Derek Bell (1980, p. 522) termed "interest convergence": that is, when the interests of dominant white groups (in these cases, their interests in enjoying and learning from cultures other than their own and/or being seen as 'inclusive') converge with the interests of racialised groups (seeing their own cultures, practices and knowledges represented in dominant institutions). However, in Connie's and Maria's stories interest convergence was combined with Leong's (2013, p. 2254) "racial capitalism"; when a racialised group is used to socially or economically benefit an otherwise white, or predominantly white institution, with little or no benefit to that group. Superficially inclusive everyday science learning activities were a double-edged sword. They placed such narrow limits on participants that

Connie was all too aware that she was not welcome or represented for the other 11 months of the year outside Black History Month. Maria and her friends felt they had been taken advantage of, not only because of the way the event had been organised, but because they realised so many objects from Latin America were in that museum but, as far as they could see, the rich science-related cultural history of Latin American people was erased.

As many have argued, the representation of cultures, knowledge and people, and who gets to have an opinion on these representations, reflect deep-seated assumptions about power (Bhabha, 1994; Gilroy, 1993; Hall, 2013; Spivak, 1988, 1999). Thus, the feelings that Connie, Maria and the other participants described can be understood as the result of structural inequalities – in particular, as relations of racial dominance, colonialism, class discrimination and oppression reproduced through forms of cultural imperialism embedded in everyday science learning practices. Practices rooted in cultural imperialism put participants in a horrible position. Should they engage with everyday science learning practices that may offer valuable forms of cultural capital, which may become particularly appealing when their own cultures, practices and knowledges are represented, however briefly? Or, do they avoid a system where even when they are repre-sented, that representation is rarely on their terms? Either way, they were left feeling profoundly uncomfortable.

Powerlessness

Powerlessness was the other key feature of participants' views about exclusion from everyday science learning, both in relation to 'race'/ethnicity and its inter-sections with class/income. Indeed, for Young (1988, 1990a), powerlessness was closely tied to class, particularly for groups with limited political or work-based authority, and who are not respected for their opinions or status. Here, I argue that experiences of powerlessness reflected the kaleidoscopic intersec-tional relationships between participants' 'race'/ethnicity and class, such that it is almost impossible to discuss these as separate issues (Gunaratnam, 2015; Puwar, 2004). For instance, Fatimata from the Sierra Leonean group argued that politically oriented everyday science learning practices such as policy con-sultations left people from racialised groups powerless and voiceless. As she angrily stated;

> We, Black people, normally think if you asked me if I would like to be part of whatever discussion based on science, to talk to the government for them to listen, we'll always say, well, they're not going to listen to us obviously, because we're minority people.

For Fatimata, participation was pointless, since her exclusion was predetermined and embedded in racist structural inequalities that shaped whose voices were heard and whose were not.

Fatimata's perspective speaks to the ideological and practical shortcomings of politically oriented science communication and public engagement practices that do not address structural inequalities. Though such practices have been critiqued from many perspectives, social justice is rarely one of them (Irwin, Jensen, & Jones, 2012; Smallman, 2016; Stilgoe, Lock, & Wilsdon, 2014). For Fatimata and the other participants attempts to engage publics with socio-scientific issues and policies were more than hollow; they reproduced social disadvantages.

When talking about exclusion, Ibrahim from the Sierra Leonean group also blamed racist practices embedded in everyday science learning activities for his exclusion. Specifically, he stated that communities like his were excluded in ways he could not influence because everyday science learning practitioners and institutions did not care about his community. That is, such practices were racist through omission. Ibrahim's point here echoes arguments made by Fatima from the Somali group and Abdou from the Sierra Leonean group in Chapter Four; they felt people from their communities were left out of everyday science learning and they did not see this as accidental. In other words, they saw that racialised groups were not part of the public that everyday science learning was made for.

The intersections of class/income and 'race'/ethnicity highlight a second way in which powerlessness affected participation. Just as leisure activities are marked by assumptions about disposable time and income, so too were everyday science learning practices (Bourdieu, 1984; Coleman & Kohn, 2007). Precarious employment was a feature of all participants' lives, entangled in their migratory trajectories and status as racialised groups in the UK (Roberts, 2004; Sassen, 2001). Their relative poverty meant certain activities were simply impossible. Participants had neither the 'free' time nor disposable income to pursue certain kinds of everyday science learning practices.

Participants' employment was often badly paid and ad hoc (see the Appendix for more details). Many participants worked around the clock with little autonomy. As Luis Diego from the Latin American group explained, exploitative work conditions and everyday science learning did not go together:

> For me it's difficult to do it [everyday science learning], because I'm working [. . .] all the time. Normally in the week I sometimes see my wife only one hour, two hours, and I have to take care of my baby; when she goes to work, I stay home. It's very, very difficult.

For participants, the money and time required to take part in everyday science learning was simply not available, rendering such practices inaccessible in ways that were beyond their control.

This finding is notable, since it belies the idea that free entry to everyday science learning activities makes them financially accessible. Entry costs have been positioned as one of the key factors that prevent certain groups from being involved in everyday science learning, particular those in science museums and science centres (Ipsos MORI, 2003; Newman, McLean, & Urquhart, 2005). As

I discussed in Chapter One, participants' relative poverty was a salient feature of their non-participation in and exclusion from everyday science learning and of course entry costs play a role. The point about free time suggests however that the economics of participation are more complex than entry costs alone.

The entrance fees of a number of British museums were withdrawn through significant government subsidy by the New Labour government of the 1990s (Ipsos MORI, 2003; Mendoza, 2017). As a result, several large museums in London, including the Natural History Museum and the Science Museum, are free to enter.[2] Interestingly, data about the kinds of visitors who visited these subsidised museums before and after entry fees were removed supports the idea that we need to think in more complex terms about costs, class, ethnicity and migration. While the number of people visiting these 'free' museums increased significantly, it turned out that this was simply because more of the same kinds of people (white, middle-class, urban families) visited these museums and repeated their visits more often (Dawson, 2014; Ipsos MORI, 2003). In other words, getting rid of upfront entrance costs did little to change the visitor profile to these museums. The economics of participation run deeper than entry costs and are about far more than socio-economic position or class background.

To understand how 'race'/ethnicity, migration, class, poverty and powerlessness play out in the economics of participation we must think about how people's choices are constrained by structural inequalities in ways they may be able to do little about. The classed notion of leisure time was, for participants, strongly inflected with issues of migration and 'race'/ethnicity, in ways that make treating these issues as analytically distinct somewhat false. For instance, in describing the competing priorities of participants' lives, Kemetta, a gatekeeper for the Latin American group, put the interplay of these issues bluntly into perspective. She told me:

> What can you do? You're actually waiting for exploitation. If you're working, you're only working for businesses who take cash in hand, it's not negotiable. I know, it's a bit heavy (laughs) so no, they're not going to museums, no lofty expectations I imaging (laughing), I know, and for them, it's [everyday science learning] not in their interests, they've not got the time.

Kemetta's comments are significant. She described how participants faced exploitation at work, where jobs were insecure, badly paid and carried out under the threat of underlying, racist debates about deportation and which people are "legal" or "illegal".

In contemporary Britain the concepts of leisure and cultural, political and educational participation are still marked by 'race'/ethnicity, class and gender (Bennett et al., 2009; Department for Culture Media and Sport, 2016; Gayo-Cal, 2006). But structural inequalities are rarely discussed in policy or practice (Belfiore, 2018). In each community, group participants described exploitative working conditions and shift-work patterns as nurses, cleaners, shop workers, or security guards, alongside family commitments, with little time for anything else. For example, Kirin, from the Asian group, explained to me that women she knew could never use everyday science learning resources: "some people, their reach is not so much, they're busy bringing up children, and life is rush, rush". Similarly, Maria from the Latin American group argued that museum visits were practically impossible because she worked back-to-back cleaning shifts, saying, "but honestly, I'm not joking, you realise the impact of working round the clock, it's so dreadful". The impact of shift-work patterns and family commitments was felt by many participants to restrict their free time such that it was described as a luxury.

From this perspective Young's (1990a) concept of powerlessness is particularly pertinent. Participants were in a position where their authority was extremely limited. As a result of their marginalisation, many options were closed to them, before the question of an entry fee could even be broached. As such, by alluding to the elite or "lofty" status of museums, Kemetta's comment quoted above highlights how unthinkable involvement in certain everyday science learning practices was for participants.

How participants imagined publics for everyday science learning

In talking about their exclusion from everyday science learning, participants described the kinds of bodies or publics they imagined *were* involved in everyday science learning practices. Participants expected publics for everyday science learning to be predominantly white. This imagined public was also shaped by participants' perceptions of free time and money associated with middle and upper classes. Notably, the kind of people participants imagined were involved in everyday science learning closely echoed the research reviewed in Chapter Two. That is, they were right to assume everyday science learning was for people from white, middle- and upper-class backgrounds. Thus, what Puwar (2004, p. 8) called the "somatic norm" of institutional practices – the kinds of bodies these practices are designed for, expect and reward – closely mirrored the kinds of publics participants expected to watch science documentaries or attend science festivals. In other words, as Gayatri Spivak (1999) has argued, European bodies are taken for granted in cultural practices (as a somatic norm), muscling out and/or obscuring alternative representations. Participants knew all too well that everyday science learning practices and their publics were made for people other than themselves and their communities. As such, both the imagined and real publics of everyday science learning disciplined participants' bodies and behaviours by situating participants as Other (Young, 1990b).

Not all bodies enjoy the same affordances in all spaces (Paechter, 2007). Thus, as discussed in Chapter Four, regardless of how much Ibrahim or Fatima liked and were interested in science, their bodies marked them out as exceptions from the somatic norm in everyday science learning. The classed, institutional whiteness of everyday science learning was particularly salient for participants. Staff, content and the other visitors, users or audiences were all expected to be white, rich and to reflect whiteness and middle/upper class values. For instance, Thomas from the Sierra Leonean group told me everyday science learning practices, on television and in museums were for "upper, middle classes. Not even just African culture, but working classes in general too, are not really involved in the culture of it". 'Race'/ethnicity and class were the key features participants used to distinguish between themselves and the imagined publics and institutions of everyday science learning.

The institutional whiteness, middle/upper class nature of everyday science learning participants described should come as no surprise. It echoes patterns of privilege and discrimination found in studies of education, museums and culture more widely (Ahmed, 2012; Bhopal, 2018; Hage, 1998; Puwar, 2004; Ware & Back, 2002). If we think about staff for instance, data from the Arts Council of England (2018) put the percentage of Black, Asian and minority ethnic staff in museums at 2.7%. Institutional whiteness is oppressive because, as Bhopal (2018, p. 25) has argued "in such white spaces, whiteness and white Western practices are the norm and those which do not comply with these are seen as outsiders and others".

While these features of everyday science learning might not come as a surprise, they worked powerfully to exclude participants. Despite the semblance of accessibility provided by the idea that anyone might turn on the television or pick up a popular science book without being stopped on the basis of their 'race'/ethnicity, class or gender, participants felt and were excluded from everyday science learning. As Connie from the Afro-Caribbean group explained, "everyone thinks the door is open, but it's not really, and that's probably because the people in charge are quite comfortable and don't want criticism or to have to change".

Scholars have rightly argued that we should be careful with claims that people *must* be involved in everyday science learning. As Susanna Hornig Priest (2009) has argued, not everyone wants to pursue everyday science learning activities. What we have to keep in mind, I suggest, is whether everyone has an equitable choice about whether they want to pursue everyday science learning (see also Barnett, Burningham, Walker, & Cass, 2010). Drawing on their perceptions of classed, institutional whiteness, participants saw everyday science learning as fields of unwelcoming, hostile practices and institutions, where they did not belong.

Imagined museums and embodied exclusion

I turn now to discuss participants' experiences and expectations of science museums and how they imagined such spaces and their publics to be. Science museums are an interesting field of everyday science learning to focus on for two reasons. First, because science museums were the most recognised, visible and, in many ways, most prestigious form of everyday science learning practice to participants. And second, because despite (or, as I argue in Chapter Three, because of) their visibility and prestige, few participants had visited anywhere like a science museum[3] before and any such visits were infrequent. As a result, how participants understood and imagined places like science museums is particularly useful for exploring tensions in how non-participation and exclusion were experienced by participants.

Despite science museums' or other informal science learning institutions' comparatively high profile compared to other forms of everyday science learning, in every group, participants were hazy about the specifics of these facilities:– where they were, what they were for, or that in London many were free to enter. Science museums were not necessarily discussed in the same breath as science centres, zoos, aquaria, botanic gardens or other informal science learning environments. Instead specific, famous London institutions were framed together as high-brow or prestigious places to visit; thus Kew Gardens (a botanic garden), the Zoological Society of London (ZSL, typically referred to as London Zoo), and the Natural History Museum (often called the 'National History Museum' by participants) were discussed together and were often conflated with other famous tourist attractions, such as the London Eye (a large Ferris wheel on the river Thames) or Madame Tussauds (a display of wax work models of celebrities).

Positioning the large London science museums as prestigious institutions and/or tourist attractions contributed to participants' views of science museums as expensive and of limited value to them, their families or communities. For example, for Maria, from the Latin American group, there was little apparent difference between the Science Museum in London and the London Dungeon, a theme park she knew was prohibitively expensive for a family of six. Although they recognised science museums as a highly symbolic form of everyday science learning, participants shared a sense of bemusement about the purpose, location and content of science museums and similar institutions. Their associated practice of not visiting science museums or similar institutions across all five groups suggests a common habitus, a taken for granted sense of alienation from these kinds of institutions (Bourdieu, 1998).

Bourdieu argued that habitus was a product of social position that "retranslates the intrinsic and relational characteristics of a position into a unitary lifestyle" (1998, p. 8). Given what I have discussed so far in this book about the characteristics of participants' positions in relation to everyday science learning as well as to science, it is perhaps not surprising that participants were not avid science museum visitors. As we saw in Chapter Three, everyday science learning operated as a series of restricted fields of cultural production, while, as discussed in Chapter Four, participants were disposed against science in general. Not visiting science museums can be seen as part of these same patterns. Some participants took not visiting science museums for granted. It was simply not part of their lifestyle. For others however, not visiting science museums arose from clear expectations about their exclusion, their sense of being out of place and their embodied discomfort.

Taking exclusion for granted

For some participants in all five groups visiting a science museum was simply unthinkable, in ways that, to them, did not seem particularly remarkable, but rather were utterly mundane. For instance, in the extract that follows, Kirin and Sarasa from the Asian group told me how close Sarasa lived to a particular science museum without ever having visited:

SARASA: I'm living nearby there, but I've never, I haven't seen the place, you pass it by

KIRIN: So how long have you lived near there?

SARASA: Ten years . . . still I haven't seen the place.

EMILY: What about the park, do you ever go in the park?

SARASA: It's a lovely park, I can see it by from the car, when we pass by.

KIRIN: Oh yeah, so you're very near to it then.

This extract highlights the everyday, seemingly natural process of how science museums and science centres (or informal science learning spaces) operate as a restricted field of cultural production and how this influences people's dispositions. Sarasa knew exactly where this particular museum was and lived nearby. She did not visit because she never had, nor did she intend to. It was simply something she did not do, an unquestioned taste for not visiting museums. For Sarasa, habitus created the conditions for "an individual agent's practices, without either explicit reason or signifying intent, to be none the less 'sensible' and 'reasonable'" (Bourdieu, 1977, p. 79). Sarasa had absorbed her exclusion from the field of science museums into her habitus in an alarmingly good example of Bourdieu's concept of symbolic violence.

The configuration of science museum visits as unthinkable, both in general and for particular institutions, was common among participants from all five community groups. For some participants their exclusion was emblematic of Bourdieu's (1990, p. 127) "gentle, invisible violence, unrecognised as such, chosen as much as undergone". Thus, Beatriz from the Latin American group laughingly described never visiting local museums, let alone larger London science museums. Similarly Idyl, from the Somali group, joked about walking into a local museum by accident, mistaking it for the library, before quickly leaving. For many participants exclusion from and non-participation in science museums was unremarkable; it was just a fact of life.

Symbolic violence can be traced therefore in how some participants did not see their exclusion from science museums as extraordinary; rather it felt natural, normal and mundane. As a result, for some participants, the inaccessibility of science museums and their exclusion from such practices was unquestioned, an accepted part of life. Sarasa did not describe an active dislike of this or other science museums, nor was she confused about how to visit it; she simply did not visit. To be clear, participants felt the weight of their exclusion in ways that were taken for granted, so some participants simply held no expectation of ever visiting anywhere like a science museum.

Expecting exclusion from science museums

Not all participants described their exclusion from science museums as unremarkable or taken for granted however. For some, although their exclusion from science museums was expected, it was not framed as natural or inevitable. Instead, these participants understood their exclusion from science museums in terms of structural inequalities, through a deeply ingrained sense that they did not belong in such spaces.

Three different conversations with participants from the Sierra Leonean group provide a useful illustration of how participants had learnt to anticipate their exclusion from science museums. In the extract that follows, Thomas from the Sierra Leonean group described his views about science museums by talking through his notion of an imagined science museum that does not fit with his own body, culture, usual behaviours, or knowledge:

> Well in a way it's like, imagine going to a museum, you don't imagine it being very comfortable, and going and what makes you comfortable, like maybe going in a hoody and just . . . people think that you have to go there a certain way, or do a certain thing when you get there or maybe like, it doesn't fit in with your culture kind of thing, you know what I'm saying? Like, could you go there with a Supermalt and just chill and just look at, you know what I'm saying? Maybe people think you have to go there and behave a certain way or have a certain amount of background knowledge already, so maybe that holds people back.

Thomas's imagined science museum was unwelcoming to people like him. From this perspective, bodies matter, not just in terms of how racialised, classed or gendered structural inequalities are projected onto bodies, but in terms of how bodies are or are not at home in certain spaces (Ahmed, 2012; Johnson, 2017; Puwar, 2004). Thus, as Yi-Fu Tuan (1977, p. 184) argued "the feel of a place is registered in one's muscles and bones" and, as participants' accounts of everyday science learning testify, these embodied feelings matter and can have lasting effects.

Thomas's anxieties were projected into his imagined science museum. Note his uncertainty about what might constitute appropriate conduct that "you have to go there and behave a certain way or have a certain amount of background knowledge". Thomas was aware that he did not have what Bourdieu called "the feel for the game" (1998, p. 80). Thomas was concerned about the differences between himself and the kinds of bodies such spaces catered for. His description highlights these differences in terms of his embodied habitus (having the wrong clothing, food, ways of walking, talking and being) and lacking the prerequisite cultural capital (his concerns about background knowledge) to successfully navigate a science museum (Bourdieu, 1990).

In the extract quoted previously, Thomas drew on his sense of the differences between 'high' and 'low' cultural practices, tastes, 'race'/ethnicity and class in terms of the forms of knowledge, behaviours and bodies that he believed would be more or less valued within a science museum (Bennett et al., 2009; Bourdieu & Darbel, 1991). As Bev Skeggs has argued, "cultures are valued differently depending on who can deploy them as a resource" (2004, p. 173). Thomas's description of a science museum highlights his expectation that his culture, knowledge and practices will not be useful to him in such a place and may instead work against him. Thomas's sense of discomfort stems from the discord between how he imagines science museums and his embodied sense of self as someone who, within such a space, is marked by 'race'/ethnicity and class. In other words, as a Black man in his early 20s, Thomas knew that he embodied what science museum visitors were not and that, as a result, his body, culture, knowledge and practices would be out of place in a science museum.

In another conversation, his friend Fatimata (also from the Sierra Leonean group and in her late 20s) repeated Thomas's point unmistakably clearly:

> If you ask me, "Fatimata are you going to come down [to a science centre or museum]", I'll tell you "No, there's not going to be anyone like me there, what's wrong with you" (laughter from Fatimata and group), but that's my perception of how I think about things.

For Fatimata visiting a science museum was simply unthinkable. Fatimata's imagined museum was populated entirely by people who were not like her, to the extent that she could not imagine visiting. From her perspective, visiting science museums was understood as wrong because "there's not going to be anyone like me there".

A final extract from a third conversation with Sierra Leonean participants emphasises even more clearly how issues of 'race'/ethnicity and class marked how museums were perceived. The extract that follows is taken from a conversation I had with Abdou, where he related a conversation he had with Ibrahim and Thomas over the previous weekend about this research project and their expectations of science museums:

> But he's [Ibrahim] never been to the science museum because he didn't even know that one exists, and I said well you see, if you had known that there's a science museum then you would have gone there in the first place but you didn't even know, so that shows that we are not catered for, and catered for meaning people from minority ethnic groups [. . .] and the reception you get it's mainly for upper class and you know, middle class people.

Abdou's description of science museums is riddled with exclusion. Not only were his friends and community excluded on the basis of 'race'/ethnicity ("we are not catered for, and catered for meaning people from minority ethnic groups"), but they were also excluded in terms of their working-class, migrant backgrounds, as exemplified by his story about Ibrahim. These extracts from Sierra Leonean participants form an explicit critique of science museums on the basis of the gap between the kinds of people participants imagined in such spaces and participants' own bodies and practices.

Thomas, Fatimata and Abdou felt science museums and spaces like them, were marked by structural inequalities – particularly racism and class discrimination – in ways that made their own exclusion inevitable. They *knew* they did not belong in science museums. As space and place geographers have long since argued, there are no neutral spaces (Massey, 1994; Tuan, 1977). Institutions can appear to be open and accessible, while their practices are actually configured such that only certain kinds of people will ever feel welcome. Thus, as Puwar (2004, p. 8) has argued, "Some bodies are deemed as having the right to belong, while others are marked out as trespassers, who are, in accordance with how both spaces and bodies are imagined (politically, historically and conceptually), circumscribed as being 'out of place'".

Exclusion from science museums was both experienced and anticipated by participants on the basis of perceived differences between their embodied selves and such spaces. These spaces, both real and imagined, disciplined participants' bodies and choices, situating them firmly as people who did not belong in science museums (Jennings & Jones-Rizzi, 2017; Young, 1990b). In a way that is distinct from the taken for granted exclusion of symbolic violence, participants *felt* excluded from science museums (Bourdieu, 1990). Thus, participants were excluded from *and* chose not to visit science museums where they knew they did not belong.

Racism, exclusion and non-participation

Participants' exclusion from science museums was a pernicious combination of a restricted field of cultural production (exclusive practices in science museums) and their resulting choice not to participate (rejecting science museums). This combination of exclusion and non-participation was produced through tensions between science museums (both real and imagined) that participants understood as white and middle/upper class spaces and their own bodies. Participants' experiences of exclusion as arising from structural inequalities rooted in racism varied within and between groups. Some felt nebulous, others concrete. Yet all came back to feelings of embodied, racialised difference. As Gunaratnam (2013, p. 48) has argued, "Some racisms you can know and name for what they are. Others are more slippery and shape-shifting, wriggling in and out of substance and language". It was experiences of racism, both vague and explicit, and their intersections with class discrimination and sexism, that were the cornerstones of participants' exclusion from everyday science learning.

Participants anticipated their exclusion from science museums such that their non-participation was ingrained in their habitus and, at times, both felt like a choice and *was* a choice. It is important then to think carefully about the complex relationships between choices, accessible opportunities, available resources, interests and feeling welcome. On the one hand participants talked about everyday science learning as a set of racist, classed and sexist practices that they felt were sufficiently unwelcoming at best and, at worst, so oppressive that they rejected the system. On the other hand, the material conditions of participants' lives, their exclusion from the fields involved in everyday science learning and the extent to which they had internalised that exclusion suggests that their involvement in everyday science learning was almost impossible. Furthermore, these features of exclusion and non-participation appeared to work together. That is, participants both could not and would not take part in everyday science learning activities. This combination of exclusion and non-participation is pernicious since it creates a resilient system that reproduces social inequalities rather than disrupting them.

To borrow again from Bourdieu, structural inequalities can be understood in terms of habitus when they create embodied dispositions around particular spaces, practices or knowledges. For people like Sarasa and Thomas, whose experiences are similarly marked by structural inequalities, in this case, experiences of racism and the intersections of racism with class discrimination, a shared habitus can develop. Thus, as Bourdieu put it, "the homogeneity of habitus is what – within the limits of the group of agents possessing the schemes (of production and interpretation) implied in their production – causes practices and works to be immediately intelligible and foreseeable, and hence taken for granted" (1977, p. 80). In other words, participants shared experiences created ways of understanding the world and acting within it that, for them, were as natural as breathing.

Feeling comfortable and at ease in a particular space can require work (Ahmed, 2000; Johnson, 2017). For participants like Sarasa the work involved in visiting a science museum was the work of trying something new, something unthinkable and subtly off-putting. For participants with more clearly defined negative experiences and expectations of science museums, like Fatimata, or Hawa and Lucille discussed earlier in this chapter, this work was extremely emotionally loaded and connected to a sense of the on-going exhaustion of living with racialised, classed and gendered discrimination and their intersections. For these participants it was as though no amount of extra work on their part could ever make them feel comfortable in a science museum. For all participants, the subtle or explicit emotional labour associated with visiting science museums was an unwanted additional workload (Ahmed, 2012; Hochschild, 1979; Lorde, 1984; Scott, 2009). From this perspective, walking into a science museum or similar everyday science learning practice carries a significant emotional burden, a burden that plays into dispositions and tastes, as a preference *not* to be in such spaces.

Summary

This chapter explored participants' perspectives on their exclusion from and non-participation in everyday science learning. Participants framed everyday science learning in terms of the cultural imperialism, powerlessness and alienation they felt. They knew all too well that they *did not fit* in everyday science learning. Participants imagined publics for everyday science learning that matched the kinds of bodies' research tells us use these resources: white, upper/middle class bodies. I focused on participants' experiences and expectations of science museums as an example of a field of everyday science learning they both felt excluded from and, in turn, rejected. Throughout the chapter I have argued that practices of racialised, classed and gendered discrimination worked to exclude participants from everyday science learning, producing a visceral, embodied sense of alienation for participants. Such practices led participants to reject everyday science learning in taken for granted and explicit ways, even at the same time as they were excluded. Thus, as far as everyday science learning was concerned instead of drawing on a sense of rightful presence, participants' habitus was marked by expectations of absence and discomfort.

That exclusion from everyday science learning is embodied in participants' dispositions against visiting places like science museums is a useful thing to learn. It means we need to think carefully about what participants might need in order to find everyday science learning more appealing. It suggests that 'quick fixes' and solutions designed around institutional interests to pick 'low hanging fruit' cannot and will not make a dent in how racism, class discrimination, sexism and their intersections structure everyday science learning. Instead, as I discuss in more detail in Chapter Seven, the data and argument discussed in this chapter suggest we must significantly disrupt how everyday science learning is practiced if we hope to convince non-participants and excluded groups that everyday science learning has some value for them.

Notes

1 I must repeat here a point made earlier that there are of course more forms of oppression at play than just 'race'/ethnicity, class, gender and their intersections. In focusing on these I do not mean to minimise issues of ability/disability, sexuality, age or other structural inequalities that may affect how people use (or not use) everyday science learning resources. However, these were not issues that emerged in this particular project, though they are documented elsewhere (for examples from museum studies see Cassidy, Lock, & Voss, 2016; Levin, 2010; Sandell, Dodd, & Garland-Thomson, 2010).

2 Although subsidised through public taxes via its central funding from government, the Science Museum in London, part of the National Museums of Science and Industry Group, put cash tills back into their entrance hall in 2014. These new tills spanned the width of the hall, such that in order to get more than a few meters into the museum, you had to walk through this new barrier. The tills are staffed and preceded by signs asking for £5.00, with, in smaller text, words about this being a voluntary donation. As a colleague told me, these 'donation' tills were part of a fundraising strategy that had been sufficiently successful that the museum was unlikely to ever remove them. Thus, the 'free' nature of these institutions is questionable in many ways, not least when high-profile museums choose to masquerade voluntary donations as entry fees.

3 Although it is not the focus of this book, I should add that participants did not report visiting other kinds of museums or art galleries either during the fieldwork for this research.

References

Ahmed, S. (2000). *Strange encounters: Embodied others in post-coloniality*. London: Routledge.

Ahmed, S. (2012). *On being included: Racism and diversity in institutional life*. Durham and London: Duke University Press.

Ahmed, S. (2017). *Living a feminist life*. Durham and London: Duke University Press.

Arts Council of England. (2018). *Equality, diversity and the creative case: A data report, 2016–17*. Manchester: Arts Council of England.

Barnett, J., Burningham, K., Walker, G., & Cass, N. (2010). Imagined publics and engagement around renewable energy technologies in the UK. *Public Understanding of Science, 21*(1), 36–50.

Bassel, L., & Emejulu, A. (2017). *Minority women and austerity: Survival and resistence in France and Britain*. Bristol: Polity Press.

Beck, U. (1995). *Ecological enlightenment*. Amherst, NY: Humanities Press.

Belfiore, E. (2018). Whose cultural value? Representation, power and creative industries. *International Journal of Cultural Policy, Online First*.

Bell, D. A. (1980). Brown v. Board of Education and the interest-convergence dilemma. *Harvard Law Review, 93*(3), 518–533.

Bell, D. A. (1992). *Faces at the bottom of the well: The permanence of racism*. New York: Basic Books.

Bennett, T., Savage, M., Silva, E., Warde, A., Gayo-Cal, M., & Wright, D. (2009). *Culture, class, distinction*. Abingdon and New York: Routledge.

Bhabha, H. K. (1994). *The location of culture*. Abingdon and New York: Routledge.

Bhopal, K. (2018). *White privilege: The myth of a post-racial society*. Bristol: Polity Press.

Bourdieu, P. (1977). *Outline of a theory of practice* (R. Nice, Trans.). Cambridge: Cambridge University Press.

Bourdieu, P. (1984). *Distinction: A social critique of the judgement of taste* (R. Nice, Trans.). London: Routledge.

Bourdieu, P. (1990). *The logic of practice* (R. Nice, Trans.). Stanford: Stanford University Press.

Bourdieu, P. (1998). *Practical reason*. Cambridge: Polity Press.

Bourdieu, P., & Darbel, A. (1991). *The love of art: European art museums and their public*. Oxford: Polity Press.

Bourdieu, P., & Passeron, J.-C. (1990). *Reproduction in education, society and culture* (R. Nice, Trans., 2nd ed.). London, Newbury Park, CA and New Delhi: Sage.

Cassidy, A., Lock, S. J., & Voss, G. (2016). Sexual nature? (Re)presenting sexuality and science in the museum. *Science as Culture, 25*(2), 214–238. doi:10.1080/09505431.2015.1120284

Coleman, S., & Kohn, T. (2007). The discipline of leisure: Taking play seriously. In S. Coleman & T. Kohn (Eds.), *The discipline of leisure: Embodying cultures of "recreation"* (pp. 1–22). New York and Oxford: Berghahn Books.

Crenshaw, K. (1991). Mapping the margins: Intersectionality, identity politics, and violence against women of color. *Stanford Law Review, 43*, 1241–1299.

Dawson, E. (2014). Reframing social exclusion from science communication: Moving away from "barriers" towards a more complex perspective. *Journal of Science Communication, 13*(2), 1–5.

Department for Culture Media and Sport. (2016). *Taking part: Longitudinal report 2016.* London: Department for Culture, Media & Sport.

de Saille, S. (2014). Dis-inviting the unruly public. *Science as Culture, 24*(1), 99–107. doi:10.1080/09505431.2014.986323

DeWitt, J., & Osborne, J. (2010). Recollections of exhibits: Stimulated-recall interviews with primary school children about science centre visits. *International Journal of Science Education, 32*(10), 1365–1388.

Falk, J., & Gillespie, K. L. (2009). Investigating the role of emotion in science center visitor learning. *Visitor Studies, 12*(2), 112–132.

Fanon, F. (2008/1967). *Black skin, white masks* (C. L. Markmann, Trans., 2008 ed.). London: Pluto Press.

Fraser, N. (2003). Social justice in the age of identity politics: Redistribution, recognition, and participation. In N. Fraser & A. Honneth (Eds.), *Redistribution or recognition? A political-philosophical exchange* (pp. 7–109). London and New York: Verso.

Gayo-Cal, M. (2006). Leisure and participation in Britain. *Cultural Trends, 15*(2), 175–192.

Gilroy, P. (1993). *The Black Atlantic: Modernity and double consciousness.* London: Verso.

Gilroy, P. (2002). *There ain't no Black in the Union Jack* (2nd ed.). Abingdon: Routledge.

Gunaratnam, Y. (2013). *Death and the migrant: Bodies, borders and care.* London: Bloomsbury.

Gunaratnam, Y. (2015). *Intersectional pain: What I've learned from hospices and feminism of colour.* Retrieved from https://www.opendemocracy.net/transformation/yasmin-gunaratnam/intersectional-pain-what-i've-learned-from-hospices-and-feminism-of-colour.

Hage, G. (1998). *White nation: Fantasies of White supremacy in a multicultural society.* New York: Routledge.

Hall, S. (2013). The spectacle of the "Other". In S. Hall, J. Evans, & S. Nixon (Eds.), *Representation* (2nd ed., pp. 215–287). London and New Delhi: Sage.

Harris, C. I. (1993). Whiteness and property. *Harvard Law Review, 106*(8), 1707–1791.

Hill Collins, P., & Bilge, S. (2016). *Intersectionality.* Cambridge: Polity Press.

Hochschild, A. R. (1979). Emotion work, feeling rules, and social structure. *American Journal of Sociology, 85*(3), 551–575. doi:10.1086/227049

Holland, D., Skinner, D., Lachiotte Jr., W., & Cain, C. (2001). *Identity and agency in cultural worlds.* Cambridge, MA and London: Harvard University Press.

hooks, b. (1981). *Ain't I a woman: Black women and feminism.* Boston: South End Press.

Hornig Priest, S. (2009). Reinterpreting the audience for media messages about science. In R. Holliman, E. Whitelegg, E. Scanlon, S. Smidt, & J. Thomas (Eds.), *Investigating science communication in the information age* (pp. 224–236). Oxford and New York: Oxford University Press.

Ipsos MORI. (2003). *The impact of free entry to museums.* London: Ipsos MORI.

Irwin, A., Jensen, T. E., & Jones, K. E. (2012). The good, the bad and the perfect: Criticizing engagement practice. *Social Studies of Science, 43*(1), 118–135. doi:10.1177/0306312712462461

Jennings, G., & Jones-Rizzi, J. (2017). Museums, white privilege and diversity: A systematic perspective. *Dimensions*, 63–74.

Johnson, A. (2017). Getting comfortable to feel at home: Clothing practices of black muslim women in Britain. *Gender, Place & Culture*, 24(2), 274–287. doi:10.1080/0966369X.2017.1298571

Kendall, M. (2015). Foreword. In *Beyond the Pale: White women, racism and history* (pp. ix–xiv). London and New York: Verson.

Leong, N. (2013). Racial capitalism. *Harvard Law Review*, 126(8), 2153–2226.

Levin, A. K. (Ed.). (2010). *Gender, sexuality and museums*. London and New York: Routledge.

Lorde, A. (1984). *Sister outsider*. Berkeley: Crossing Press.

Massey, D. (1994). *Space, place and gender*. Cambridge: Polity Press.

McIntosh, P. (1989, July/August). White privilege: Unpacking the invisible knapsack. *Peace and Freedom*, pp. 10–12.

McIntyre, A. (1997). *Making meaning of whiteness: Exploring racial identity and white teachers*. New York: University of New York.

Mendoza, N. (2017). *The Mendoza Review: An independent review of museums in England*. London: Department for Digital, Culture, Media & Sport.

Miles, A., & Gibson, L. (2016). Everyday participation and cultural value. *Cultural Trends*, 25(3), 151–157. doi.org/10.1080/09548963.2016.1204043

Moraga, C., & Anzaldúa, G. (2015/1981). Introduction, 1981. In C. Moraga & G. Anzaldúa (Eds.), *This bridge called my back*. Albany: State University of New York Press.

Newman, A., McLean, F., & Urquhart, G. (2005). Museums and the active citizen: Tackling the problems of social exclusion. *Citizenship Studies*, 9(1), 41–57. doi:10.1080/1362102042000325351

Paechter, C. (2007). *Being boys, being girls: Learning masculinities and femininities*. Maidenhead: Open University Press.

Puwar, N. (2004). *Space invaders: Race, gender and bodies out of place*. Oxford and New York: Berg.

Roberts, K. (2004). Leisure inequalities, class divisions and social exclusion in present-day Britain. *Cultural Trends*, 13(2), 57–71. Retrieved from www.informaworld.com/10.1080/0954896042000267152

Saha, A. (2018). *Race and the cultural industries*. Cambridge: Polity Press.

Sandell, R., Dodd, J., & Garland-Thomson, R. (2010). *Re-presenting disability: Activism and agency in the museum*. Abingdon and New York: Routledge.

Sassen, S. (2001). *The global city: New York, London, Tokyo* (2nd ed.). Princeton, NJ and Oxford: Princeton University Press.

Scott, S. (2009). *Making sense of everyday life*. Cambridge: Polity Press.

Skeggs, B. (2004). *Class, self, culture*. London and New York: Routledge.

Smallman, M. (2016). Public understanding of science in turbulent times III: Deficit to dialogue, champions to critics. *Public Understanding of Science*, 25(2), 186–197. doi:10.1177/0963662514549141

Spivak, G. C. (1988). Can the subaltern speak? In C. Nelson & L. Grossbery (Eds.), *Marxism and the interpretation of culture* (pp. 272–313). Chicago: University of Illinois Press.

Spivak, G. C. (1999). *A critique of postcolonial reason: Toward a history of the vanishing present*. Cambridge, MA and London: Harvard University Press.

Stilgoe, J., Lock, S. J., & Wilsdon, J. (2014). Why should we promote public engagement with science? *Public Understanding of Science, 23*(1), 4–15.

Tuan, Y.-F. (1977). *Space and place: The perspective of experience.* Minneapolis, MN: University of Minnesota Press.

Ware, V., & Back, L. (2002). *Out of whiteness: Color, politics, and culture.* Chicago and London: University of Chicago Press.

Young, I. M. (1988). Five faces of oppression. *The Philosophical Forum, 19*(4), 270–290.

Young, I. M. (1990a). *Justice and the politics of difference.* Princeton, NJ: Princeton University Press.

Young, I. M. (1990b). *Throwing like a girl and other essays in feminist philosophy and social theory.* Bloomington, IN: Indiana University Press.

Chapter 6

Being excluded

As a former museum employee, I knew it was as easy as looking across a crowded gallery to know science museums were not very inclusive. Surprisingly, at that time (the mid-2000s), little research was available about how exclusion from everyday science learning worked, let alone how such practices could become more inclusive. My colleagues and I found that job titles like 'Community Officer' and 'Diversity Manager' were sometimes used by our institutions to partition equity issues off from day-to-day work, often (though not always) against the best intentions of those involved (Ahmed, 2012; Jennings & Jones-Rizzi, 2017; Tlili, 2008). These mounting frustrations were part of why I wanted to explore equity and exclusion in everyday science learning from a different perspective, the perspective of those who did not, would not or could not participate.

Excluded by design

In this chapter I discuss participants' experiences of everyday science learning in specific practices of their choice. As part of the research for this book, participants from four of the five community groups chose to do an everyday science learning activity with me. Three groups (the Asian, Latin American and Sierra Leonean groups) chose to visit science museums with exhibitions of natural history objects, 'hands-on' science exhibits, and exhibits of living plants and animals such as gardens, butterfly houses and aquaria. The Somali group decided to visit an interactive science centre with computer-based exhibits and audio-visual 'science shows' about various scientific phenomena.

I borrowed the idea of accompanied visits as a research practice from the field of Museum Studies (see for example Hooper-Greenhill, Moussouri, Hawthorne, & Riley, 2001). It is a form of ethnographic participant observation, where researchers join participants in an activity (Brewer, 2000; Hammersley & Atkinson, 1997). In this case the accompanied visits were not positioned as interventions with measurable outcomes or as comparative studies, but rather as tangible experiences for participants and me to engage with together. I simply joined in with their visits, travelling with participants to the everyday science learning activity they wanted to do, going through the experience with them, joining them for lunches or snacks and travelling home with them afterwards. In each case I covered costs for transport, food and refreshments.

Three different sites were visited because the Sierra Leonean and Asian groups independently decided to visit the same museum. Visits lasted between 2 and 5 hours. Notably, with echoes of Sarasa's comments discussed in Chapter Five, the Afro-Caribbean group chose not to visit or take part in any everyday science learning activities as part of the study. They told me that they were happy to talk, but as Irene put it, they were simply not interested enough in science or everyday science learning to "go on an outing".

In this chapter I describe what happened during four visits to two different science museums and a science centre. When I initially analysed the data that this chapter is based on, I tried to apply theories about how people learn and engage with science in museums to it. Here you can see the big assumption that was driving how I thought the visits would work. I assumed visits to science museums and a science centre would support participants' engagement with science and science learning, at least in some way. I was wrong. Instead, the visits reinforced what participants already knew, that science museums and science centres were not for them, and, by association, nor was science.

The inequitable and unequal patterns of participation in everyday science learning described throughout this book, as well as participants' strongly held views that everyday science learning was not for them, were borne out in the visits. Rather than finding that visits to the science museums and the science centre were enjoyable and/or supported participants' science learning and engagement, I found exclusion was embedded in the practices participants encountered. Thus, as I discuss in what follows, from the economic costs, to issues of language, representation and staff support, participants were Othered through the everyday science learning practices they encountered. Understanding participants' experiences of exclusion in practice is crucial if, as I discuss in Chapter Seven, we are to develop meaningfully inclusive, equitable everyday science learning practices.

The economics of museum visits

Free admission was standard across all three of the institutions we visited and all participants' expenses on research days were covered. Nonetheless, participants found visits expensive in both explicit and more subtle ways that affected their access to science museums and science centres, as I discuss in this section. There was a widespread assumption across all five community groups that informal science learning institutions like museums charged prohibitively expensive entrance fees. For example, Abdou, from the Sierra Leonean group, described feeling very anxious about entry costs:

> Do you have to pay to go to these museums though? 'Cos that's one thing I always think, that you have to pay, you have to book, ok how can I go there

I haven't got a credit card, because maybe they'll not accept the debit card I've got, you know.

For Abdou, this was a litany of off-putting access problems. His concerns were underscored by twin anxieties about personal finances and the process of paying. Abdou, like participants in every group, expected science museums to be expensive.

Notable in the four visits was the extent to which expectations of high costs were met by what participants encountered in the museums and the science centre. Participants from each community group noted the additional costs associated with the visit, such as the expense of travelling to these spaces, the price of items in their shops, and the high cost of food and drink. Although participants' socio-economic positions had been disrupted in various ways by their different migration trajectories (as discussed in the Appendix), they were not rich in the UK. Instead, precarious employment conditions, unemployment and relative poverty were shared features of their lives, against which science museums and centres were seen as incredibly rich and incredibly expensive institutions. The institutions we visited were in large, beautiful buildings, buildings that were part prestigious, historic monument, part snazzy scientific laboratory. The stark differences in economic capital between the science museums and the science centre on one hand and the participants on the other was extremely visible to participants (Bourdieu & Darbel, 1991; Macleod, 2005). Two of the Latin American participants, for example, expressed distaste at the sight of donation boxes filled with money. These displays of wealth did double duty, as a request for even more money from an evidently rich institution, they signalled to participants how little they belonged.

For participants, the difference between their economic capital and the expenses involved in a museum visit was significant; they added up and confirmed their expectations that science museums were not for them. Thus, for Sofia and Flor, sisters from the Latin American group, the effects of the expense of museum visits on visitor demographics was unmistakable after their visit:

SOFIA: People with a higher income as well would be more likely to go,
FLOR: What do you mean?
SOFIA: 'Cos having a trip out for a day costs a lot more money than you kind of think. Even if it's say, like, initially free to get in, you're talking travel, food costs, going into the gift shop, all of that. So I think for people who've got higher incomes it's not really an issue, whereas for other people, "oh well, there's however many of us, that's going to add up"

Sofia understood the hidden costs of visiting a science museum as a series of signs telling her the institution was not for her. She and her sister clearly echo the comments made by Thomas, Fatimata and Abdou from the Sierra Leonean group discussed in Chapter Five; in economic terms, science museums were marked by cost as not for them. As Bourdieu and Wacquant put it, "when habitus encounters a social world of which it is the product, it is like a 'fish in water': it does not feel the weight of the water, and it takes the world about itself for granted" (1992, p. 127). Participants in each group felt the "weight of the water" in the science museums and the science centre even without entry fees, the high costs of visiting were all too apparent.

Visits were also associated with an opportunity cost by participants. As discussed in Chapters Three and Five, participants had little or no leisure time with which to visit science museums or science centres. Thus, in terms of economic capital, not only was money needed to access a museum, but it played a fundamental role in whether participants had the time to visit in the first place (Bourdieu & Wacquant, 1992). This meant that the visits we went on were seen as highly unusual and difficult to repeat because of the relationship between science museums and participants' social positions.

The structure of the science museum field, in which possession of time and disposable income are normalised preconditions for successful participation in museum or science centre visits, was not without injury for participants. For example, Maria from the Latin American group, raised concerns about visits to science museums in our conversations following her visit:

> I feel guilty that I'm not doing it all the time. . . . Nothing can beat an outing, but you do want to have the cash in your pocket, it's bad enough the little ones saying "I need this", and as well as that everyone else on the street listening.

Maria framed her guilt in terms of being unable to provide "outings" for her family and a sense of shame at not having enough "cash" in her pocket. Implicit within Maria's statement is a critique of the costs associated with science museums. The situation is complicated however, since Maria explicitly blamed herself, rather than the structures and practices of science museums, for not being able to take her family to visit such spaces. Complex and negative feelings lie at the heart of symbolic violence. Thus Bourdieu argued "symbolic violence accomplishes itself through an act of cognition and of misrecognition" (1992, pp. 171–172), resulting in personalised feelings of guilt rather than placing the blame for structural inequalities on institution, systems or society.

From this perspective, the economics of participation in everyday science learning, and science museums or science centres more specifically, is clear. These practices were marked by racialised, class discrimination, such that without free time and disposable income visiting was both impossible and unthinkable for participants. There is a question of redistributive social justice here; science museums and science centres were not accessible Participants' experiences of science museums and science centre practice confirmed for them that the somatic norm of such spaces was that of upper- and middle-class bodies.

The language of exclusion: literacies and inaccessibility

Participants found that being able to use everyday science learning practices required more than physical access. In this section I discuss how participants struggled to navigate the science museums and the science centre we visited in ways that made them feel at all comfortable, let alone supported science engagement and learning. Literacy, in its most straightforward form – being able to speak and read a language – was a key access issue for certain participants from each community group during the visits. All three institutions relied exclusively on English as their institutional language, with no evidence of materials in alternative languages apparent during the visits. Participants were, as a result, faced with what Bourdieu described as "the imposition of the dominant language and culture as legitimate and by the rejection of all other languages into indignity" (1998, p. 46). Institutional reliance on the English language to the apparent exclusion of all other languages worked as a form of cultural imperialism (Young, 1990). Not being totally fluent in English left participants unable to access science learning opportunities and profoundly uncomfortable in the spaces we visited. In this section I discuss how English literacy issues compounded problems of scientific literacy and museum literacy for participants as a result of the language and interpretation choices made by the institutions we visited.

On multiple occasions participants translated or attempted to translate exhibit text, staff speech, or museum signs for one another, as other studies have found (Ash, 2004). Within-group translation could not compensate however, for the language barriers participants encountered. At a fundamental level, this emerged when some participants were unable to read signs or exhibit texts, as happened in every group. In practice, participants' own languages and cultures were invisible in the science museums and the science centre in favour of English text and European (often nationalistic, British) stories. While participants were recruited on the basis that we could communicate with one another in English, this did not mean all participants were fluent readers of English. English was a second, third, fourth or fifth language for all participants. As a result, literacy issues significantly affected participants' visits, restricting their access to scientific information, science learning activities and opportunities to build or leverage capital. Furthermore, as I argue throughout this section, monolingual institutional language practices in combination with the assumptions of scientific literacy, museum literacy and other content literacies embedded in the spaces we visited situated participants as racialised Others and exacerbated their exclusion and marginalisation.

Inaccessible science learning

Institutional reliance on monolingual English language practices created significant problems for participants' access to science learning opportunities. At the most basic level exclusive institutional use of English meant participants were all too often blocked out of exhibits. During the Latin American group's visit, for example, Ignacio tried and failed to help his daughters use interactive computer exhibits. Audio and text instructions were highly detailed, delivered simultaneously, included scientific terminology and were always only in English. Ignacio was left to rely on his bilingual daughters for translation and unable to help them in return.

Participants found that along with English literacy, scientific literacy was frequently a prerequisite for making sense of and being able to use an activity or exhibit. Scientific literacy is often described as the cornerstone of science education (Kelly, 2010). It is, however, understood in varied and contested ways (Brown, 2006; Norris & Phillips, 2003; Nowotny, Scott, & Gibbons, 2003). From a Bourdieusian perspective for instance, being able to understand and use scientific language such as 'hypothesis' or 'nucleus' could, for example, be considered as both linguistic and cultural capital. Whatever side of these debates you take, scientific literacy – whether linguistic, skills based, focused on applications or politics – can be seen as a form of science capital (Archer, Dawson, DeWitt, Seakins, & Wong, 2015). As other studies have found, however, the science museum practices participants encountered were designed such that participants were expected to walk into an exhibit with a significant amount of scientific knowledge to successfully use exhibits (Bain & Ellenbogen, 2002; Schlereth, 1992; Tunnicliffe & Laterveer-de Beer, 2002).

Exclusive practices around English literacy and scientific literacy were reinforced by other assumptions built into institutional practices that made learning science especially difficult for participants. In terms of museum literacy – the ability to understand how to use or look at an exhibit – the rules of interaction for science museum and science centre exhibits were complex, especially for interactive exhibits. A traditionally displayed exhibit of an object in a glass box, aquarium or animal enclosure meant participants struggled to read about the object but could still enjoy looking at it. In contrast, interactive exhibits with moving pieces or exhibits based on computers posed serious challenges. Interactive exhibits were complicated both *logistically* (hitting the right part of a touch screen, aligning your body correctly for a camera) and *conceptually* (tasks often follow a teaching pattern of 'elicit action' and 'response from interactive', typically based on understanding scientific content).

Being able to use a museum exhibit, especially an interactive exhibit, often means following specific steps in order (Heath, Lehn, & Osborne, 2005). In this sense, a degree of museum literacy was assumed in the design of museum practices. As Sharon Macdonald has argued, following Stuart Hall, exhibits are encoded with meaning, which is more or less accessible, more or less understandable, depending on how visitors decode them (Hall, 1980; Macdonald, 1998). Thus, even knowing how to look at an exhibit is not straightforward (Bourdieu & Darbel, 1991; Rice, 1992). Furthermore, knowing how to look at or use an exhibit has racialised, classed and gendered nuances. For instance, in their review of museum and science centre practices in the US, Marilyn Fenichel and Heidi Schweingruber (2010, p. 120) concluded that everyday science learning institutions "often privilege the science-related practices of middle-class whites and may fail to recognise the science-related practices associated with individuals from other groups" (p. 120). Similarly, Duensing (2006) found that the assumptions made about how exhibits work in science museums had a cultural component. That is, understanding an exhibit may involve different steps in Brazil than in Poland. Notably, for participants, being able to make sense of, use and learn science from the exhibits they encountered during the visits required culturally specific forms of cultural capital (English literacy, museum, literacy, scientific literacy and other content literacies such as European geography or history). Thus assumptions made about the kinds of people using these spaces – the institutional somatic norm – resulted in practices that Othered and excluded participants.

A useful example of how participants did not fit with the somatic norm assumed by the institutions we visited comes from data from the Somali group's visit to a science centre. The overlap of institutional language practices (based on particular assumptions about English literacy) with expectations about the knowledge visitors brought with them (based on assumptions of scientific and museum literacy) created a learning opportunity that was impossible for the Somali participants to access despite their best efforts. The extract that follows is from Hamiido, attempting to use one of the interactive computer exhibits with me. Hamiido was in her mid-30s, had two smart phones and had previously shown me how to use a couple of apps on my phone. She was an adept user of technology in many ways, but that did not help her in the science centre. I joined her at this exhibit, having witnessed her difficult first attempt:

(*A voice in an upper class, southern English accent from the machine talks again and says "Now it's over to you, remember, you need to keep the conditions right, the humidity, the pH . . ."*)

HAMIIDO: How can I know?
EMILY: I think it's going to tell us?
HAMIIDO: Yeah?
EMILY: – Plant cells are growing – so look.
HAMIIDO: This is coming up.
EMILY: Yeah, pH is going down, so, so the temperatures, ok, oh down with the temperature.
HAMIIDO: This one?
EMILY: Yep.
HAMIIDO: Ok.
EMILY: Oh, temperature too low, up a bit, ok, pH low, pH up . . .
HAMIIDO: Oh, confusing.
EMILY: I know.
HAMIIDO: Yes, this is for scientists, yeah?

To use this exhibit Hamiido needed first to understand spoken and written instructions in English at the same time (with no way to slow down or repeat instructions). She then needed to understand several scientific terms and concepts *and* she needed enough museum literacy to understand how exhibits like this one might work.

Let's break it down. In terms of scientific literacy, the previous extract shows Hamiido needed to know what plant cells were and how they grew, the effect of pH, temperature, water, humidity, waste and nutrients. In terms of museum literacy, she also had a lot to do. This interactive computer exhibit appeared to be designed for multiple users, with a large screen and several moving components of image and text, which proved difficult to use alone or in a pair. To make things worse, a loud siren went off at regular intervals as part of the experience. Thus, as Doris Ash and Judith Lombana (2011, p. 3) found working with people from linguistic minorities in informal science learning institutions in the US "they often needed to do extra work to 'figure out' what the exhibit 'wants' them to do". If we take a step back from this episode, we can start to see how even with extra work, it was almost impossible for Hamiido to benefit from using this exhibit. As Hamiido discovered to her detriment, despite working hard to use this exhibit, she literally could not play the game.

Bourdieu and Johnson argued that access to the artistic field of cultural production was restricted, "accessible only to those who possess practical or theoretical mastery of a refined code, of successive codes, and of the code of these codes" (1993, p. 120). In addition to the inaccessibility of everyday science learning I discussed in Chapter Three, I suggest a similar practice of restricted access operated *within* the practices of the science centre and science museums we visited. For instance, in the previous example, the learning opportunity designed by the science centre implicitly assumed visitors understood multiple codes, not least scientific and museum literacy, premised on an overarching assumption of English fluency.

Here Frantz Fanon (2008/1967) is helpful in framing institutional language practices as rooted in colonialism. The science on display was European science, peopled with European names, faces and stories. Knowledges, practices or images of other people were absent, whitewashed or presented as the subjects of European scientific discovery. Colonialism and cultural imperialism were rooted into these institutions (Young, 1990). Thus, in these spaces, museum language and scientific language *was* the English language. As a result, for participants the epistemic practices of the museums and the science centre reflected and re-created colonial practices of knowledge, power and racism that served to exclude and marginalise them. In other words, these museums did not help participants to engage with, enjoy or learn about science. Instead, participants learnt science was not by or for then.

Inaccessible institutions

Institutional practices that assumed various literacies were not just a problem for engaging with or learning about science; they undermined participants' whole visits. For instance, in terms of access, the science museums and science centre were designed with text, and therefore English reading ability, as the cornerstone of way-finding signs, cafés, shops and toilets, as well as exhibit materials. Although some museums and other informal science learning institutions in the UK do provide some translation to other European languages, we did not see any during the visits. As a result, monolingual institutional language practices made participants feel uncomfortable, unwelcome and unwanted, situating them as racialised groups in white British institutions.

Participants' visit experiences show how language worked as a form of symbolic institutional capital for the science museums and the science centre, restricting access not only to science learning, but to whole institutions (Bourdieu, 1991). It signalled to participants that they were out of place, defined in "relation to language" as Other in these spaces (Bourdieu & Passeron, 1990, p. 116). Because participants did not fit the somatic norm of the science centre or science museums we visited they had to do a considerable amount of additional labour to navigate these spaces, getting confused and sometimes upset. For example, some Sierra Leonean participants struggled to read exhibit texts but also felt anxious and lost because they could not follow institutional signage and did not feel confident asking for directions in English. Mama Kamara and Mama Sesay, elders from the Sierra Leonean group, concluded for instance, at the end of their visit that they would not feel comfortable returning to the museum. As they put it:

MAMA KAMARA: The only problem we have, because, lot of we, some of the people they learn a bit, but some of we no went to school, don't know anything, so if you only take bus to come and see these people here, maybe going inside, getting lost, because you don't know, you don't know how to read, to this way, exit, this that? You can't know, so unless you come with somebody we know, then you come and see the place.

MAMA SESAY: Because if you come back, you might not want to ask, if you want to be sure, you need to ask somebody, which my way, I want to go out, or I want to go up, or I want to see something and they will show you 'go this way', and when if you want to go out and you don't know, you can ask someone, I want to go out, I'm tired.

Without being able to read signs or ask for directions, Mama Kamara and Mama Sesay felt not only that the whole museum was inaccessible for them, but that they were themselves out of place, anxious, unsure of what to do and unable to return. Indeed, the symbolic violence at work here is significant (Bourdieu & Passeron, 1990). The institutional reliance on English undermined their visits and left Mama Kamara and Mama Sesay with a sense of not being sufficiently educated, linguistically deficient, and profoundly out of place in the museum (Bhopal, 2018).

While language-based exclusion may in some senses appear an obvious and taken-for-granted aspect of access to everyday science learning, it was a key problem for participants. As Idyl from of the Somali group noted during an interview a few weeks after our visit, staff at science centre claimed it was "open to everyone". But, as she pointed out, the institution did not cater for the linguistic practices of local minority ethnic communities. From her perspective, not representing global languages such as Arabic or French symbolised a closed and exclusively white institution. She could recognise the somatic norm of these institutions and how different it was from her and her community. For Idyl, the purportedly accessible science centre was, in practice, linguistically structured in ways that made it almost impossible for visitors with limited English language fluency to use. Thus, like other studies, participants found they needed to be able to speak and, almost more crucially, to be able to read the language of the institution they visited, in order to understand not only exhibit texts, but the signage, opening hours, toilets, café menus and interactions with staff (Ash, 2004; Garibay, 2009; Tenenbaum & Callanan, 2008; Yalowitz, Garibay, Renner, & Plaza, 2013).

The somatic norm evident in participants' visits had much in common with that discussed in Chapter Five, and unpacking issues of literacy and forms of capital helps to show how assumptions about museum visitors and, as a result, institutional practices, are marked by 'race'/ethnicity. Monolingual language practices shore up practices of institutional whiteness. British publics are still by and large imagined as white, despite considerable evidence to the contrary in cities across the country (Gilroy, 2002; Modood, 2010; Puwar, 2004; Ware & Back, 2002). From a literacies perspective, English language practices can be understood to operate within these science museums and the science centre as a racialised form of cultural capital, reinforcing which publics matter and which do not (Fanon, 2008/1967; Hage, 1998; Puwar, 2009). Marked by language practices as racialised Others, participants had to work hard just to be in these spaces, knowing they were bodies out of place, before accessing science learning activities even came into the picture. Notably, the burden for making the visits work was on their shoulders, not on the institutions'.

The extra work created by being Othered in specific everyday science learning practices extends therefore beyond the emotional burden discussed in Chapter Five, into the work involved in navigating spaces where you do not fit. For participants this problem was (at least) two-fold. In terms of access, institutional use of English blocked participants from science learning opportunities and left them unsupported in terms of navigating the museum or science centre as a whole (Dawson, 2014; Rawls, 1971). But these already oppressive blocking practices reflect racist assumptions, as Fanon (2008/1967) has argued, premised in this case on English, as emblematic of whiteness and knowledge. Thus we must also consider whose bodies, practices and knowledges are recognised, represented and respected (Fraser, 2003; Young, 1990). For instance, as Kathleen Yep (2014) has shown, deciding to present English-only texts, signage, marketing and so on is a political choice and a choice that sends a strong message to people from racialised groups about who is and is not valued. Such practices reinforce the cultural imperialism participants already associated with everyday science learning. That is, they were Othered by the epistemic practices of these spaces. Literacies then, in their various forms, can be seen as key issues for equity and exclusion in everyday science learning in museums and science centres, both in terms of access and representation.

Encounters with staff: more hindrance than help

The staff participants met in the science museums and the science centre we visited were not necessarily the supportive presence they might have been. Staff facilitation has been found to make or break a visit (Ash & Lombana, 2011; Rahm & Ash, 2008; Ruiz-Funes, 2008; Uyen Tran & King, 2007). The staff that visitors meet can provide valuable connections to the institution, help visitors use exhibits, engage with scientific content or just find the café. As such staff were the face and the body of the institution, in a way that was as important, if not more so, than the other visitors participants met. Frustratingly, participants' interactions with staff echoed the other exclusive institutional practices they encountered in two key ways. First, participants' encounters with café staff, shop staff and security guards were key to feeling welcome and, at times, emphasised how out of place they felt. Second, experiences of educator-led exhibitions or workshops reinforced participants' perceptions that science museums, science centres and science were not for them.

Interactions with museum and science centre staff who were not in education roles played a notable part in participants' visits. The Sierra Leonean group were delighted to find a security guard from Sierra Leone at work in the museum they visited. He went out of his way to welcome the group to the gallery he was working in and his presence was remarked on repeatedly by participants after the visit. They were surprised, but happy, to see someone from Sierra Leone in the museum, but noted their surprise emerged in response to the whiteness of other staff and visitors. In sharp contrast, in what might be best described as acts of racial profiling, participants from the Latin American group were closely watched and repeatedly asked by security guards and exhibition facilitators not to touch the animals in the special exhibition we bought tickets for. Maria eventually became furious; neither she nor her daughters touched the animals at all and no-one else was being singled out by staff. Staying with the Latin American group, the family were upset again after being asked by café staff and security guards to leave well in advance of the museum's closing time and well in advance of other visitors.

These encounters provides good examples of what Lucille from the Sierra Leonean group described as the "look" that security guards sometimes gave her in museums, expensive shops or airports to make her feel unwelcome. Being the subject of additional scrutiny because their bodies did not look like the somatic norm expected in the science museums and the science centre we visited happened to participants in every group. Although in one gallery, for the Sierra

Leoneans, this was positive, in general it was not. For Fanon (2008/1967) and Puwar (2004) the 'look' that Lucille described is a racialised one, one that marks the person who is looked at as Other and out of place. As Puwar (2004, p. 49) writes, "This is a visibility that comes from not being the norm. It is a process that is not all that different from the way in which racialised groups are visible on the street, and especially in particular locations heavily demarcated as white places". In this respect certain kinds of encounters with staff seemed premised on suspicion for participants' out of place bodies, bodies that were highly visible because of their difference from the somatic norm of the spaces we visited. The surveillance, not to mention the more vocal disciplining of participants' behaviours and bodies by guards, floor and café staff, made participants feel profoundly uncomfortable in ways we should not underestimate when thinking about equity, exclusion and everyday science learning.

Participants' encounters with educators *also* reinforced how out of place participants felt. The two visits with experiences facilitated by staff trained in inclusive, community education practices were particularly revealing in how institutional practices about what who counts are deeply embedded, even in practices designed to support access to everyday science learning for minoritised groups. The Somali group visit to the science centre was booked in advance, so the centre knew they were meeting an all-female Somali group. The staff member assigned to the group for the facilitated session with the interactive computer exhibits was a young man, who introduced himself as coming from a South Asian background and growing up in a poor neighbourhood in a large city in the north of England. He described how he had "used education and science to sort of escape, 'cos I used to be on quite a rough council estate". While introducing the session and the work of the science centre, Deepak drew at length on his sense of science identity (Carlone & Johnson, 2007). He was evangelical about the power of science education to change lives, telling participants that he worked "directly with young people, trying to sort of inspire them with science". He used identity politics to position himself at the start of the session, outlining how he was similar to the participants (disadvantaged neighbourhood, racialised minority background) but also different to them (had a science degree, worked in a science centre and, notably for the Muslim Somali women, male).

As the visit continued, Deepak struggled to create a comfortable environment for the Somali participants. He asked few questions about participants, but instead posed question after question about the content of the science centre and scientific information that they could not answer. Deepak's explanations of the interactive computer exhibits used what Bourdieu (1990, p. 108) described as the "status authority" conferred upon him as a science centre facilitator. They were fast and left little space for participants to ask questions or understand what his instructions meant, as the extract shows:

DEEPAK: Have you played top trumps before? The card game? Have you? No?
IDYL: No.

DEEPAK: All right so this is a two player game, so you see you've got these catego-
ries, you need to pick a number that's going to be higher than hers.

IDYL: Ok.

DEEPAK: So if the number 10 would be higher, you'd pick the card, but if you
think your number four is going to be higher than hers, like, for example,
carbon dioxide is one of the main air pollutants, so just press on that, so it
actually turns out that there's a lot more nitrogen in the air than carbon
dioxide, so she wins the card, and it's your turn again (he says to Nadifa).

NADIFA: Oooh.

DEEPAK: Did you get all that, so go higher than hers, so try not to lean on it
remember.

IDYL: Ok.

DEEPAK: So you need to lift up your other hand, because you're leaning on it, so
pick a number that's going to be higher than hers, you've trumped 700,000,
her number's a lot bigger, she has 20 million, so you win the card, 'cos it's
winner stays on

The talk in this extract was dominated by the staff member, who raced through
his explanation, pausing only to tell Idyl off for leaning on the screen. It does not
come across on the page, but when I listened back to the recording, I was shocked
to hear that Deepak shouted most of his instructions. He created few spaces for
the Somali participants with their limited English fluency to interact with him
beyond the question "did you get all that?" Deepak's explanation assumed scientific literacy from participants, such as the differences between atmospheric gases
and their familiarity with a British children's game, top trumps. While Deepak
did not appear to "feel the weight of the water" during the visit, seeming fully
immersed in how this field worked, the Somali participants did (Bourdieu &
Wacquant, 1992, p. 127). Idyl, for example, later commented on how confusing
she found the interactive exhibits, saying "I didn't know what I was doing; I was
just touching, sometimes I was winning really without knowing the reason why".
This example shows how institutional practices and unspoken assumptions about
somatic norms created extra work for participants *and* left them feeling Othered
and unwelcome.

The Asian group had a similar experience during an hour-long, object-handling
workshop with a community educator from the museum we visited. Unlike the
Somali participants, the Asian participants were positive about their workshop
experience and the facilitator they worked with. I too was impressed with this
workshop. It was certainly one of the better ones across the visits, representing
what could be considered good practice across the field. The facilitator worked
hard to learn participants' names, tried to link the handling collection to their
everyday lives and supported them to ask their own questions.

Rather than building a bridge between participants, the institution and science however, participants' reflections suggest instead that, much like the Somali
group's experience, the workshop re-created a sense that science museums and

science were not for participants, as this extract from my conversation with Kirin three weeks after the visit shows:

KIRIN: The workshop, workshop, was wonderful I think; the thing that most fascinated me was the elephant tooth, so big and heavy, I can't imagine that you know. I still remember her saying, that animal thing, is it a cat that people wear for good luck? On their shoulders, with the face [. . .]

EMILY: So what was it about that workshop that made it stand out, that made it seem special to you?

KIRIN: Yes, the people who were trying to show us everything I think were wonderful, very intelligent people, they have interest in these things, and it was it was different from everyday life, what things we see.

Notable in the extract is how Kirin differentiated between herself and the museum staff. She emphasised the differences between herself, her community and their daily lives versus the unusual, special objects, and intelligent staff at the museum. This particular staff-led experience was positive in several ways for Kirin, but a sense of social distance outweighed any sense of building bridging social capital with new people or disrupting the power relationships between herself and the museum (Putnam & Goss, 2002). For Kirin there was a significant distance between herself and museums as "wonderful", staffed by "very intelligent people". Similar comments were made by others from the Asian group who had been on the visit. Thus building on the discussions in Chapters Three and Four, this session did nothing to support Kirin to feel that science was for her, but rather fed into her disposition against science and everyday science learning.

Participants found that even educational practices intended to be inclusive work to reify distinctions between the somatic norm of science museums and science centres on one hand and participants on the other. The two facilitated sessions were focused first on scientific content and institutional practices and, despite being led by experienced educators, did not focus on participants' backgrounds, assets or needs. As discussed in Chapter Two, designing museum and science centre experiences that prioritise the needs of science and the needs of institutions over and above the needs of participants are unlikely to lead to inclusive, equitable practices. Instead, participants' encounters with museum and science centre staff confirmed how out of place they were in these spaces.

Not designed for us

As might be expected given the visit experiences discussed here, participants in these four groups felt no more inclined to adopt a practice of visiting science museums or science centres after the visits than they did before. As Idyl ultimately concluded in the weeks following her visit, science museums and science centres were "not designed for us". Participants were not blind to the somatic norm of the institutions they visited. As well as noticing the differences between themselves and most of the staff they saw, participants also commented on not seeing other visitors like themselves. As Maria, from the Latin American group explained, retelling her daughter Flor's comments after the visit:

> she was saying "look at all the people going in, mum, they're all white, middle class and wealthy, aren't they, there's not many people I was at college with here". She made that observation and a lot of it is the money thing, being able to afford it, even though supposedly it's free.

In pinpointing the 'race'/ethnicity and class status of the museum visitors, Maria and Flor spoke to the somatic norm of the science museum they visited. They knew they did not fit in. As their dad, Ignacio, asked me, weeks later "for example, we went to the butterflies museum, I didn't see even one or two Latin families there; did you see anywhere, or no?" Their take home message had been that science museums and science centres were not for them.

The racialised nature of exclusive institutional practices was particularly striking. For participants involved in the visits, their differences from the somatic norm were easily seen in not only the faces of other visitors, staff and the images in exhibitions, but in whose knowledge was on display, whose practices were reified, whose food was sold in cafés and whose languages formed the backbone of these institutions. Crucially, we can read this list in terms of McIntosh's (1989) work on white privilege, as institutional practices organised around an assumed somatic norm – white, middle/upper class bodies with English, scientific and museum literacies – that worked to exclude, discomfort and Other participants. As a white, middle-class woman, brought up in the UK with a science degree and a museum studies degree these spaces work for me in ways that I might misrecognise because I am so accustomed to the privilege of feeling at home in them. Participants' experiences highlight a stark contrast from my own. Rather than feeling at home they were made to feel unwelcome, incapable and Othered in ways that confirmed their pre-existing disposition against everyday science learning. Thus, despite being physically inside these buildings, they were excluded because their exclusion was embedded within institutional practices rooted in racism (Ahmed, 2012; Bhopal, 2018; Puwar, 2004).

Reflecting on their visits participants either flat out rejected the notion of returning or talked about returning to a science museum or science centre as a difficult and unlikely choice, contingent upon an unrealistic alignment of factors. As Idyl from the Somali group put it:

IDYL: Well if I was going with people that we get on, like, then maybe yes (laughs) but if you didn't have anywhere else to go, then maybe yes, just have a laugh and then pretend that we have visited somewhere today, that could be it.

EMILY: But if there were other places to go?

IDYL: Then I wouldn't put that first (laughs).

To be clear, Idyl bluntly suggested she would only return to the science centre if there was nowhere else to go in London with her friends. It was extremely unlikely. The extract suggests Idyl's habitus, in particular, her disposition against taking part in everyday science learning practices was reinforced, rather than transformed or disrupted, through her visit.

Through their visit experiences, participants across these four groups simply confirmed that museums and science centres were not designed for them. Bourdieu argued that habitus is an "open system of dispositions that is constantly subjected to experiences, and therefore constantly affected by them in a way that either reinforces or modifies its structure" (Bourdieu & Wacquant, 1992, p. 133). The visits carried out in the research for this book show that rather than changing participants' attitudes and behaviours, the visits reinforced participants' dispositions against science museums and science centres. As such, as Noah Feinstein (2017) has argued, the extra work required to visit everyday science learning spaces that neither represent nor respect your knowledges, culture or self only compound patterns of exclusion.

The visits provide insights into how power works in everyday science learning, drawn from experiences of exclusion. Participants' exclusion was embedded in the practices of the spaces they visited. In this sense, as Skeggs has argued, "institutional apparatus exists to institutionalise entitlements" (2004, p. 151). Entitlements which participants experienced in reverse through the visits, where the institutions marked them as Other and unentitled. Institutional practices of exclusion were most marked, for these participants, by structural inequalities, particularly 'race'/ethnicity and class, and can therefore be understood as practices of racism and class discrimination. By any vision of equity and inclusion, these are not acceptable or appropriate practices for any informal science learning institution or everyday science learning activity. The visits therefore highlight how certain bodies struggle to take up spaces that are not designed for them. Thus, from a social justice perspective, the experiences of the minoritised groups who took part in this project provide an important critique of informal science learning practices, and everyday science learning more broadly.

Reproducing exclusion and non-participation

The visits highlight how narrowly conceived much of the research on publics, informal science learning, science communication and everyday science learning has been. As discussed in Chapter Three, significant claims are made about the benefits of participation in everyday science learning activities of various kinds. Instead, as the research for this book shows, we need to rethink those assumptions. The science museums and the science centre we visited operated as restricted fields of cultural production (Bourdieu & Johnson, 1993). Participants were unable to use or benefit from their visits because the institutions were designed around forms of capital and literacies they did not possess. At the same time, the forms of capital and literacies participants held were not represented or valued by these institutions. As a result, participants were unable to leverage their own knowledges and practices to use, accrue or exchange forms of capital during their visits.

The cultural imperialism at work in how the experiences, knowledges and practices of dominant groups were enmeshed with the design of institutional practices made it impossible for Othered visitors to participate in informal science learning on anything like an equal footing (Young, 1990). To even be in these spaces participants had to work much harder than other visitors, not only to try to access science learning opportunities, but to be themselves in spaces that configured them as racialised, classed Others, highly visible as a result of their embodied differences from the somatic norm. As such, visiting science museums or science centres can be understood to come at a significant cost to participants who are symbolically marked as out of place by institutional practices. The cost of feeling out of place and Othered by virtue of practices that reproduce social disadvantages at the intersections of racism and class discrimination. The cost of changing who you are to fit in.

This book explores what happens when people from racialised groups encounter science through the practices that mediate between people and science in the public sphere, from a sociological perspective. If we think about this chapter in light of the rest of this book, we have to reconsider some of the underlying assumptions made about everyday science learning. At best, participants from outside the imagined public and somatic norm of everyday science learning practices did not learn science. And that's before we even get into debates about whose knowledges, practices and selves that science-related content speaks to. At worst, participants experienced marginalisation, powerlessness, cultural imperialism, racism, class discrimination and sexism based on structural inequalities that reinforced rather than ameliorated social disadvantages. This means if we are to take the challenge of developing meaningfully inclusive and equitable everyday science learning practices seriously, we must disrupt and transform these practices in ways that go far beyond community ticket schemes or outreach programmes to challenge core practices around content and media.

Summary

This chapter discussed the four accompanied visits (to science museums and a science centre) carried out as part of the research for this book. The visits testified to how little deficit oriented, assimilationist approaches to understanding equity and exclusion from everyday science learning worked from participants' perspectives. Being inside these buildings of everyday science learning did not mean participants were included. Far from it. Participants were Othered by the sheer extent to which they were made to feel different from the somatic norm of the institutions we visited. The economic costs of museum visits, the complex and overlapping literacy issues faced, issues of representation and science content and the deeply problematic interactions with staff left participants even less inclined to visit such spaces. Thus institutional practices rooted in racism and its intersections with class discrimination worked to ensure participants felt unwelcome, disrespected and alienated in these spaces.

In this chapter I argued that participants had to do a significant amount of extra work, not only to try to access science learning opportunities, but to be in these spaces at all. From a social justice perspective this is clearly an unfair burden. Participants should not have had to act as well behaved guests in science museums, who have to play by mysterious and confusing rules in order to be able to admire content and practices they can neither access nor shape. Participants should not have had to work so hard just to be in spaces where they were so evidently marked out as Other and out of place. Critically, though they were generous enough to take part in this research, they also should not have had to do the work of demonstrating how everyday science does not work. As Judit Moschkovich (2015/1981, p. 73, italics in the original) put it "We've all heard it before: *it is not the duty of the oppressed to educate the oppressor*". It is on us then, the producers, users and researchers of everyday science learning, to take seriously the experiences participants shared and to work towards developing equitable, inclusive practices, as I discuss in Chapter Seven.

References

Ahmed, S. (2012). *On being included: Racism and diversity in institutional life*. Durham and London: Duke University Press.

Archer, L., Dawson, E., DeWitt, J., Seakins, A., & Wong, B. (2015). "Science capital": A conceptual, methodological, and empirical argument for extending bourdieusian notions of capital beyond the arts. *Journal of Research in Science Teaching*, *52*, 922–948. doi.org/10.1002/tea.21227

Ash, D. (2004). Reflective scientific sense-making dialogue in two languages: The science in the dialogue and the dialogue in the science. *Science Education*, *88*(6), 855–884.

Ash, D., & Lombana, J. (2011). *Reculturing museums: Scaffolding towards equitable mediation in informal settings*. Paper presented at the National Association for Research in Science Teaching, Orlando, FL.

Bain, R., & Ellenbogen, K. M. (2002). Placing objects within disciplinary perspectives: Examples from history and science. In S. G. Paris (Ed.), *Perspectives on object-centered learning in museums* (pp. 140–155). Mahwah, NJ: Lawrence Erlbaum Associates, Inc.

Bhopal, K. (2018). *White privilege: The myth of a post-racial society*. Bristol: Polity Press.

Bourdieu, P. (1990). *The logic of practice* (R. Nice, Trans.). Stanford: Stanford University Press.

Bourdieu, P. (1991). *Language and symbolic power* (G. Raymond & M. Adamson, Trans.). Cambridge and Malden, MA: Polity Press.

Bourdieu, P. (1998). *Practical reason*. Cambridge: Polity Press.

Bourdieu, P., & Darbel, A. (1991). *The love of art: European art museums and their public*. Oxford: Polity Press.

Bourdieu, P., & Johnson, R. (1993). *The field of cultural production: Essays on art and literature*. Cambridge: Polity Press.

Bourdieu, P., & Passeron, J.-C. (1990). *Reproduction in education, society and culture* (R. Nice, Trans., 2nd ed.). London, Newbury Park, CA and New Delhi: Sage.

Bourdieu, P., & Wacquant, L. (1992). *An invitation to reflexive sociology.* Chicago: University of Chicago Press.

Brewer, J. D. (2000). *Ethnography.* Buckingham and Philadelphia: Open University Press.

Brown, B. A. (2006). "It isn't no slang that can be said about this stuff": Language, identity, and appropriating science discourse. *Journal of Research in Science Teaching, 43*, 96–126. doi.org/10.1002/tea.20096

Carlone, H. B., & Johnson, A. (2007). Understanding the science experiences of successful women of color: Science identity as an analytic lens. *Journal of Research in Science Teaching, 44*(8), 1187–1218. doi.org/10.1002/tea.20237

Dawson, E. (2014). Equity in informal science education: Developing an access and equity framework for science museums and science centres. *Studies in Science Education, 50*(2), 209–247. doi.org/10.1080/03057267.2014.957558

Duensing, S. (2006). Culture matters: Informal science centres and cultural contexts. In Z. Bekerman, N. C. Burbules, & D. Silberman Keller (Eds.), *Learning in places: The informal education reader* (pp. 183–202). New York: Peter Lang Publishing.

Fanon, F. (2008/1967). *Black skin, white masks* (C. L. Markmann, Trans., 2008 ed.). London: Pluto Press.

Feinstein, N. W. (2017). Equity and the meaning of science learning: A defining challenge for science museums. *Science Education, 101*(4), 533–538. doi:10.1002/sce.21287

Fenichel, M., & Schweingruber, H. A. (2010). *Surrounded by science: Learning science in informal environments.* Washington, DC: The National Academies Press.

Fraser, N. (2003). Social justice in the age of identity politics: Redistribution, recognition, and participation. In N. Fraser & A. Honneth (Eds.), *Redistribution or recognition? A political-philosophical exchange* (pp. 7–109). London and New York: Verso.

Garibay, C. (2009). Latinos, leisure values, and decisions: Implications for informal science learning and engagement. *The Informal Learning Review, 94*, 10–13.

Gilroy, P. (2002). *There ain't no Black in the Union Jack* (2nd ed.). Abingdon: Routledge.

Hage, G. (1998). *White nation: Fantasies of White supremacy in a multicultural society.* New York: Routledge.

Hall, S. (1980). Encoding/decoding. In S. Hall (Ed.), *Culture, media, language: Working papers in cultural studies, 1972–79* (pp. 107–116). Birmingham: Umwin Hyman (Publishers) Ltd.

Hammersley, M., & Atkinson, P. (1997). *Ethnography* (2nd ed.). London and New York: Routledge.

Heath, C., Lehn, D. V., & Osborne, J. (2005). Interaction and interactives: Collaboration and participation with computer-based exhibits. *Public Understanding of Science, 14*(1), 91–101.

Hooper-Greenhill, E., Moussouri, T., Hawthorne, E., & Riley, R. (2001). *Meaning making in Art Museums 1: Visitors' interpretive strategies at Wolverhampton Art Gallery.* Leicester: Research Centre for Museums & Galleries.

Jennings, G., & Jones-Rizzi, J. (2017). Museums, white privilege and diversity: A systematic perspective. *Dimensions,* 63–74.

Kelly, G. (2010). Scientific literacy, discourse, and epistemic practices. In C. J. Linder & L. A-Stman (Eds.), *Exploring the landscape of scientific literacy* (pp. 61–73). New York: Taylor & Francis Group.

Macdonald, S. (1998). Exhibitions of power and power of exhibition: An introduction to the politics of display. In S. Macdonald (Ed.), *The politics of display: Museums, science, culture* (pp. 1–24). New York: Routledge.

Macleod, S. (2005). Rethinking museum architecture: Towards a site-specific history of production and use. In S. Macleod (Ed.), *Reshaping museum space: Architecture, design, exhibitions* (pp. 9–25). London and New York: Routledge.

McIntosh, P. (1989, July/August). White privilege: Unpacking the invisible knapsack. *Peace and Freedom*, pp. 10–12.

Modood, T. (2010). *Still not easy being British: Struggles for a multicultural citizenship*. Stoke on Trent: Trentham Books.

Moschkovich, J. (2015/1981). But I know you, American woman. In C. Moraga & G. Anzaldúa (Eds.), *This bridge called my back* (pp. 73–77). Albany: State University of New York Press.

Norris, S. P., & Phillips, L. M. (2003). How literacy in its fundamental sense is central to scientific literacy. *Science Education*, *87*(2), 224–240. doi.org/10.1002/sce.10066

Nowotny, H., Scott, P., & Gibbons, M. (2003). Introduction: "Mode 2" revisited: The new production of knowledge. *Minerva*, *41*(3), 179–194. doi:10.1023/a:1025505528250

Putnam, R., D., & Goss, K., A. (2002). Introduction. In R. D. Putnam (Ed.), *Democracies in flux: The evolution of social capital in contemporary society* (pp. 3–20). Oxford and New York: Oxford University Press.

Puwar, N. (2004). *Space invaders: Race, gender and bodies out of place*. Oxford and New York: Berg.

Puwar, N. (2009). Sensing a post-colonial Bourdieu: An introduction. *The Sociological Review*, *57*(3), 371–384. doi:10.1111/j.1467-954X.2009.01856.x

Rahm, J., & Ash, D. (2008). Learning environments at the margin: Case studies of disenfranchised youth doing science in an aquarium and an after-school program. *Learning Environments Research*, *11*(1), 49–62. doi.org/10.1007/s10984-007-9037-9

Rawls, J. (1971). *A theory of justice*. Cambridge, MA: Harvard University Press.

Rice, D. (1992). Vision and culture: The role of musuems in visual literacy. In S. K. Nichols (Ed.), *Patterns in practice* (pp. 144–152). Washington, DC: Museum Education Roundtable.

Ruiz-Funes, C. R. (2008). Mediation within science centres and museums: The guides of Universum, México. *Journal of Science Communication*, *7*(4), 1–4.

Schlereth, T. J. (1992). *Cultural history and material culture: Everyday life, landscapes, museums*. Charlottesville: University Press of Virginia.

Skeggs, B. (2004). *Class, self, culture*. London and New York: Routledge.

Tenenbaum, H. R., & Callanan, M. A. (2008). Parents' science talk to their children in Mexican-descent families residing in the USA. *International Journal of Behavioral Development*, *32*(1), 1–12.

Tlili, A. (2008). Behind the policy mantra of the inclusive museum: Receptions of social exclusion and inclusion in museums and science centres. *Cultural Sociology*, *2*(1), 123–147.

Tunnicliffe, S. D., & Laterveer-de Beer, M. (2002). An interactive exhibition about animal skeletons: Did the visitors learn any Zoology? *Journal of Biological Education, 36*(3), 130–134.

Uyen Tran, L., & King, H. (2007). The professionalization of museum educators: The case in science museums. *Museum Management and Curatorship, 22*(2), 131–149. doi.org/10.1080/09647770701470328

Ware, V., & Back, L. (2002). *Out of whiteness: Color, politics, and culture.* Chicago and London: University of Chicago Press.

Yalowitz, S., Garibay, C., Renner, N., & Plaza, C. (2013). *Bilingual exhibit research initiative: Institutional and intergenerational experiences with bilingual exhibitions.* Retrieved from Washington, DC: Centre for Advancement of Informal Science Education.

Yep, K. S. (2014). Reimagining diversity work: Multigenerational learning, adult immigrants, and dialogical community-based learning. *Metropolitan Universities, 25*(3), 47–66.

Young, I. M. (1990). *Justice and the politics of difference.* Princeton, NJ: Princeton University Press.

Chapter 7

Transforming everyday science learning

It is easy to talk about equity, exclusion and inclusion; it is quite another thing to translate that talk into meaningful action. Think back to Connie's comment in Chapter Six: *how uncomfortable do we have to be in order to try to change established practices?* In this chapter I discuss the key implications of the research discussed in this book: that everyday science learning is exclusive and that change is needed to develop accessible, equitable practices. This chapter starts by focusing on some of the rare examples of moments where participants *were* able to connect with science during their museum visits. In the rest of the chapter I focus on how inclusion can be understood in everyday science learning and propose an access and equity framework based on the research discussed in this book.

In this chapter I outline a theory of inclusion for everyday science learning. I argue that we can think about inclusion without assuming excluded groups are deficient or that assimilation is the price of inclusion (as discussed in Chapter Two). The access and equity framework I describe combines ideas from social justice, critical pedagogy and the work of Bourdieu. The framework, for me, is a thinking tool to reimagine, evaluate and understand inclusion in everyday science learning in practice. In other words, in this chapter I build on my analysis of exclusion to discuss how access, equality and equity might be achieved in everyday science learning.

In thinking about how everyday science learning could be usefully disrupted and transformed, I think we need to radically reimagine practice. For instance, what would be involved in actively pursuing anti-racist, anti-classist, anti-sexist everyday science learning practices that were neither ableist, heteronormative or otherwise oppressive? Clearly here lies what Toni Dancstep once described to me as the "creative challenge" for those involved with everyday science learning (2018). Obviously I cannot hope to list practice-based solutions for every inclusion challenge people may face, but my hope is that the ideas discussed in this chapter provide thinking tools that support the development of more inclusive everyday science learning practices (Eisenhart, 2001).

The cynics amongst you may, at this point,[1] be asking why I think everyday science learning practices are worth putting any more energy into. Given the argument I have made throughout this book – that exclusion is built into everyday science learning practices in ways that reproduce structural inequalities – why should we try to rethink and rework them? Would we be better off starting something entirely different? I am not trying to be flippant. The oppression of minoritised groups is, in many ways, over-determined in our societies (Spivak, 1999). For me there is a serious point here, drawing on Audre Lorde (1984), about whether tools that play a part in reproducing social hierarchies can ever be sufficiently reworked to disrupt rather than re-create social disadvantages.

My answers are pragmatic ones. First, while I am very interested in pursuing alternative, new and radically different everyday science learning practices, the fact is we are faced with a landscape where certain practices, institutions and their funding streams are firmly established. But that does not mean these institutions and practices cannot be repurposed. If we can reimagine everyday science learning using ideas from redistributive social justice (so that more people can benefit from the advantages they offer) and relational social justice (so that more people, practices, knowledges and forms of capital are valued and represented), we may be able to create disruptive, transformative practices. In other words, as the activists involved in the UK's Museum Detox (2018) group put it, we may be able to 'detox' these practices.

My second answer stems from the data generated in the research for this book. Although I found exclusion was embedded within and across everyday science learning practices, a few, brief transformative moments stood out from the accompanied visits. In particular, on a few occasions participants were able to reclaim parts of their visits by drawing on their own resources to enact practices they valued in the spaces we visited and, as a result, to generate value for themselves, as I discuss in the first part of this chapter. These were clearly not the relevant, inclusive and meaningful science learning experience that might be hoped for. But for me, these moments suggest there *is* some potential for developing everyday science learning practices that are more meaningful, more inclusive and more equitable.

Dancing with natural history: trying to transform the museum

Although the museum and science centre visits carried out as part of the research described in book were, by and large, unsuccessful in terms of providing accessible or enjoyable science learning opportunities for participants, they were not wholly without any moments of science-related meaning making. Participants were occasionally successful "space invaders" in Puwar's terms (2004, p. 7). Able to use their own backgrounds to transform an exhibit into something about which they could draw meaning from, to disrupt a museum space to better suit themselves, to reflect their own knowledges and practices. These moments were brief and, as I discuss in what follows, somewhat limited. Such moments are significant, however, given the broader context of inaccessibility and marginalisation experienced by participants because they suggest everyday science learning opportunities, not least those in museums or science centres, could be reimagined in more inclusive, equitable and respectful ways.

When it was possible to do so, participants used their own knowledge, languages and cultures to connect with and make sense of science during their visits. For example, participants from the Asian group recognised fish from Bangladesh, told stories about fishing as children in other countries, and shared fish recipes in the aquarium. Similarly, Ignacio from the Latin American group, unable to use the English language resources of the museum, still facilitated learning opportunities for his daughters when he was able to do so using his own resources. He brought a display of scorpion specimens to life telling spine-chilling tales of his childhood encounters with scorpions in Colombia. These moments can be understood in pedagogic terms as instances of cross-cultural learning where different contexts and transnational experiences are pulled together and shared (Aikenhead, 2002; Mai & Ash, 2012; Rahm, 2010; Roth, 2008; Yalowitz, Garibay, Renner, & Plaza, 2013). Through these experiences, some participants were able to connect their cultural backgrounds and identities with science and everyday science learning. These moments, while rare, enabled some participants to suddenly recognise something of themselves in these institutions. For that instant they reworked the somatic norm of the space, becoming people who could relate to the exhibits, texts, images and concepts presented by the museums. In doing so, participants also reconfigured and disrupted knowledge and practices within the museum spaces (Dawson, 2014b; Puwar, 2004; Sandell, Dodd, & Garland-Thomson, 2010).

In this sense, to build on the discussion in Chapter Three, certain participants were sometimes able to use the museums or the science centre they visited to build on their own assets – or forms of capital – which may represent a powerful way for people from minoritised backgrounds to create relevant cultural and educational opportunities within everyday science learning (Archer, Dawson, Seakins, & Wong, 2016; Gonzalez, Moll, & Amanti, 2013; Trienekens, 2002). This is no small thing. Think about the context in which these moments took place. The epistemic practices of the science centre and museums we visited privileged certain forms of literacy and knowledge, as evidenced in learning opportunities where English and scientific language, skills and concepts were a prerequisite for engagement. There were no revisionist texts discussing the under-representation of people from minoritised groups, whether women, racialised groups or working-class groups. Indeed, these exclusions appeared to be quite unremarkable in the spaces we visited. Little support or recognition was given to facilitate alternative explorations of exhibits, objects, or other learning opportunities within the science centre or the museums. Thus to be able to connect with the spaces we visited on their own terms should be seen as a significant success for participants, and again, required extra labour from them.

The most striking example of cross-cultural meaning making occurred during the visit with the Sierra Leonean group. One participant recognised a bird amongst a display of several animals, which prompted some members of the group to dance and sing next to the exhibit for several minutes. Through their dancing and singing in a natural history exhibition, some of the Sierra Leonean participants briefly disrupted their disempowered relationship with the museum to create connections between themselves, their cultural heritage and particular exhibits. In the extract that follows, Hawa and Mama Kamara explain the relevance of the bird on display for their community:

HAWA: It's a song for that (she points to the bird).
MAMA KAMARA: For these two birds. When you sing, go to society, people eat for you (sings a bit of the song as an example then pauses). You know that cassava?
EMILY: Yeah
MAMA KAMARA: They boil it, and dry it, the cassava, when they dry it, that's what we call [acolopala].

Participants had performed the ceremonial dance for a rite of passage involved with hunting and eating the bird on display. The participants appeared to enjoy themselves, dancing, singing, laughing and talking. The dance can be understood as cross-cultural meaning making. It can also be understood as a form of transgression and resistance to the museum, with its objects imprisoned in glass, inaccessible texts, unknown codes of conduct, and incomprehensible languages. By dancing and singing, some of the Sierra Leonean participants briefly disrupted the quiet gallery with movement and noise, making the space their own and transforming their experience of inaccessibility and discomfort by working with an exhibit on their own terms and validating their own knowledges.

In the weeks after the visit however, some Sierra Leonean participants expressed feelings of anxiety and regret about their behaviour when talking to me about their visit to the museum. Without prompting about the dancing or singing, some participants described feeling concerned that, as Hawa later put it in an email, they had been "carried away" when they engaged with the museum in this way. For the Sierra Leonean participants, dancing and singing represented a struggle to change, however briefly, their social position as unknowledgeable and out of place in the museum. Such moments of cross-cultural meaning making are complex and not unproblematic. Although the Sierra Leonean participants had tried to find a way to be themselves in the museum, they later felt they behaved in ways that marked them as Other; as too loud, too boisterous, too African. In this sense participants once again felt the difference between their selves and the somatic norm of science museums and similar spaces (Ahmed, 2000; Johnson, 2017). Not only in embodied terms, but in terms of their practices, their knowledges, practices and their ways of being. Thus, as Bourdieu reminds us, making use of dominant systems (such as those rooted in cultural imperialism, powerlessness and other forms of oppression that exacerbate structural inequalities) is "easier for the dominant than for the dominated" (1977, p. 82). In other words, being a space invader, disrupting and transforming who and what matters in a museum setting, is far from easy.

While moments of cross-cultural learning arose during the visits, the extent to which such moments were transformative was limited, leaving some participants feeling they had, in Hawa's words, "misbehaved". Bourdieu described educational institutions, such as museums, as sites that defend "cultural orthodoxy or the sphere of legitimate culture against competing, schismatic or heretical messages" (1993, p. 112). That is, dominant cultural, educational and political practices, such as those involved in everyday science learning, actively resist the kinds of disruption and transformation the Sierra Leonean participants were able to briefly enact.

Although moments of cross-cultural meaning making suggest there is potential for transforming everyday science learning, we would be foolish to rely on them, not least since the burden of labour fell once again onto participants. Instead, as I have argued throughout this book, significant change is required before everyday science learning practices could be considered inclusive and equitable. We must rethink, reimagine and reconfigure everyday science learning practices such that access is radically opened-up, so that the knowledges, practices and selves of those currently excluded are respected, represented and valued amongst the forms of literacy and capital reproduced in these activities.

A theory of inclusion for everyday science learning

If we understand exclusion from the various fields involved in everyday science learning practices as comprised of multiple structural inequalities and their intersections, how might we understand what inclusion looks like? What does it mean to create new practices and new knowledges from those that have supported social reproduction in favour of dominant groups, and at the expense of the minoritised? From a social justice perspective, it must mean thinking in terms of disrupting and transforming such practices and knowledges, to rebuild new cultural, educational and political practices – new practices which, as Bhabha (1994) has argued, are already happening and, what's more, are inevitable.

The key implications of the research discussed in this book are that everyday science learning practices are exclusive and must become more inclusive. This part of the chapter explains how I think inclusion could be achieved. I begin by discussing a theory of inclusion in everyday science learning based on the data and theories presented in this book. I review how exclusion operates, based on the chapters in this book so far. I then outline what accessible and equitable practice involves. Here I draw on concepts from work on social justice, namely redistributive and relational social justice (Fraser, 2003; Rawls, 1971; Young, 1988, 1990, 2000). Finally, I describe how these ideas could translate into practice. To do that I draw on three key concepts (infrastructure access, literacies and community acceptance) that build on the Bourdieusian analysis of exclusion discussed in this book to think through what accessible, equitable everyday science learning could be like (Dawson, 2014a, 2017; Grabill, 1998; Porter, 1998). The access and equity framework for everyday science learning practices that I outline in the last part of this chapter is my attempt to explore what this model of inclusion might mean in practice.

Understanding exclusion

Since the majority of the book is concerned with understanding exclusion, I focus here on summarising the arguments and data discussed already. I have argued that by exploring the everyday science learning experiences of people from minoritised, and specifically, racialised groups we can develop insights into systems of exclusion and social reproduction. As I have shown through the chapters of this book, participants' exclusion from everyday science learning was complex, resulted from structural inequalities, not least the pernicious effects of the intersections of racism and class discrimination, and importantly, was not their fault.

Throughout this book I have argued that everyday science learning practices are marked by and reproduce structural inequalities. In Chapter Three I argued that the various fields involved in everyday science learning operated as restricted fields of cultural production (Bourdieu & Johnson, 1993; Miles & Gibson, 2016). That is, their inaccessibility was what gave such fields their cachet or privilege. In this sense, participants' exclusion formed a key part of what made everyday science learning special. In Chapter Four I argued that participants were disposed against science 'in general'. By looking in more detail at the lives of three specific participants, I showed that such shared dispositions – or habitus – arose as the result of structural inequalities, specifically colonialism, racism and class discrimination. As such, even those participants who tried to pursue science, whether in education or as a career, struggled to do so.

In Chapter Five I argued that participants were both excluded from and chose not to participate in everyday science learning. In this sense, they were disposed against everyday science learning as well as science. Once again, I showed how participants' dispositions were the result of everyday science learning practices rooted in racism, class discrimination and, for some, sexism. In Chapter Six, looking at the example of one specific field of everyday science learning – science museums and science centres – I argued that participants exclusion was embedded in institutional practices so that these spaces operated as restricted field of cultural production, even once participants were inside.

Participants' experiences of exclusion were affected by access to the fields of practice involved in everyday science learning, the forms of capital (the knowledges, skills, literacies and 'know-how') required to be successful in those fields, and their habitus or disposition towards (or, in this case, away from) participation in everyday science learning (Bourdieu, 1998; Bourdieu & Wacquant, 1992). The exclusive effects of field, capital and habitus compounded one another and were oppressive for participants. Thus, the exclusion of racialised groups can be seen as a structural feature of everyday science learning. The role of everyday science learning practices in social reproduction is therefore clear: such practices work to reproduce patterns of social disadvantage (for the excluded) and privilege (for the included).

Oppressive practices are all too easily reproduced when they are not seen as joined-up, systematic and active practices of oppression. People, both inside and outside everyday science learning can become too accustomed to the system, and in turn reproduce the practices they are used to, whether that's designing inaccessible activities, reifying whiteness in exhibits or not visiting museums. Thus systems are reproduced and change becomes hard to imagine, especially radical change.

The deficit model of exclusion and the crusade model of inclusion positions the problem with those who experience exclusion, not with the organisations/institutions, policies, practices and practitioners (researchers included) who create, perpetuate and benefit from exclusive practices. Throughout this book I have argued that these are not issues of deficit, to be addressed by means of an assimilationist crusade. Exclusion results from structural inequalities – racism, sexism, class discrimination (and more) and their intersections – embedded in everyday science learning practices. These exclusive practices are resilient precisely because of their systematic, accepted, almost invisible, almost deniable nature. It is perhaps too tempting to look away from the almost invisible problems. You only need to narrow your eyes a little and they disappear. Instead, as I argued in Chapters Five and Six, the extra work required to participate in everyday science learning that currently falls on minoritised groups must fall to those of us already involved in this system, whether as audiences/visitors/users, producers, funders, researchers or policy makers.

Across all these chapters questions about access, representation and respect arose again and again as interwoven issues that were frequently inseparable. It is not enough to provide access to a practice that Others, disrespects and oppresses already marginalised groups. Inclusion into unchanged everyday science learning practices means Othered publics must 'pass' as the somatic norm. This means rendering their bodies, knowledges and selves invisible whilst putting up with cultural imperialism and powerlessness that are marked by racism, sexism, class discrimination and their intersections, and thereby exacerbate their marginalisation. This is not inclusion. Instead we must transform practice, not just in terms of access, but in terms of representation too. Questions of redistributive and relational social justice must therefore be addressed *together* when trying to understand exclusion from everyday science learning and, in turn, when trying to think about what inclusion might mean.

From this perspective, developing inclusive everyday science learning practices is not a matter of tweaking a practice here or there. Instead, I argue that to become accessible and equitable, significant transformation is required across all the fields involved in everyday science learning. We need therefore to move beyond approaches that seek to leave practice-as-normal unchanged wherever possible. Such approaches cannot transform entrenched patterns of non-participation and exclusion in restricted fields of cultural production (Ahmed,

2012; Dawson, 2018; Lynch & Alberti, 2010; Puwar, 2004). Disruption and transformation are required because without them, entrenched social inequalities around participation in everyday science learning will persist.

Thinking about social justice

To think about what inclusion in everyday science learning could involve I turn here to theories of social justice. I wanted first to find a way to think about inclusion that was not premised on assuming excluded publics were ignorant or deficient. I also wanted to make sure that inclusion was not framed as assimilation. To recap from Chapter Two, social justice theorists have long been concerned with how resources might be distributed most equally (redistributive social justice) and most equitably (relational social justice). In the first model, justice is about equality of access and opportunity between social groups – that is, everybody being able to do, enjoy or use the same amounts of the same things (Fraser & Honneth, 2003; Rawls, 1971). The second model, in contrast, emphasises the value of recognising, appreciating and validating difference. In other words, recognising, respecting and valuing the fact that people differ and taking those differences into account, rather than treating everyone's needs as the same (Taylor, 1994; Young, 1990, 2000).

Of course, redistributive and relational models of social justice need not sit in opposition to each other. As Fraser (2003) has argued, combining both models of social justice provides us with a powerful tool for thinking about and addressing issues of exclusion and inclusion. This dual approach to social justice can help us to think about practices in everyday science learning along a continuum of weak to strong forms of inclusion (see Figure 7.1, below). A weak form of inclusion only addresses one facet of social justice (access/equality *or* equity), while a strong form of inclusion addresses both. For instance, if we apply these ideas to scientific practices, scholars have argued that it is not enough to recruit more ethnically diverse scientists, more female scientists or more scientists from working-class backgrounds, without simultaneously changing the culture and content of scientific knowledge (Hammonds, 2009; Harding, 2008; Longino, 1990; Schiebinger, 2007; TallBear, 2013).

Understanding inclusion from a dualist view of social justice requires a commitment to exploring beyond issues of access to specific fields and particular forms of capital (weak inclusion). It means asking questions of whose knowledge counts, whose bodies are welcome, changing forms of representation, opening up who has power and creating cultural change (strong inclusion). In terms of everyday science learning therefore, practices that support better access (physical and intellectual) and at the same time represent, respect and validate the practices and knowledges of marginalised groups are more inclusive than those which support only one or the other. In essence, in Figures 7.1 and 7.2, which follow, the top right-hand square on the grid is the sweet spot for thinking about what inclusive practice might involve.

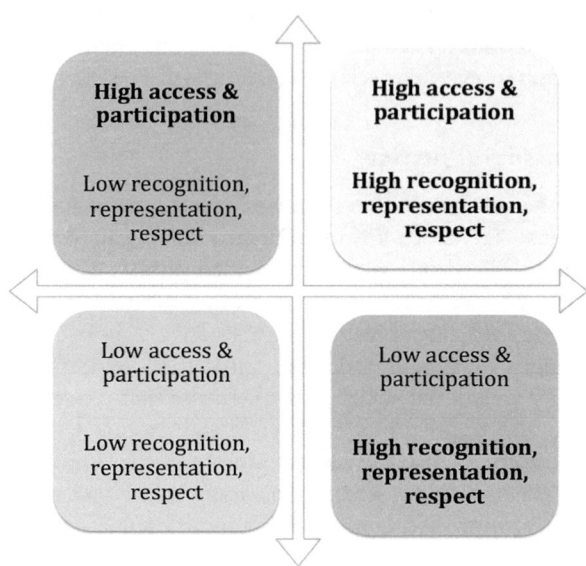

Figure 7.1 A dualist model of redistributive and relational social justice

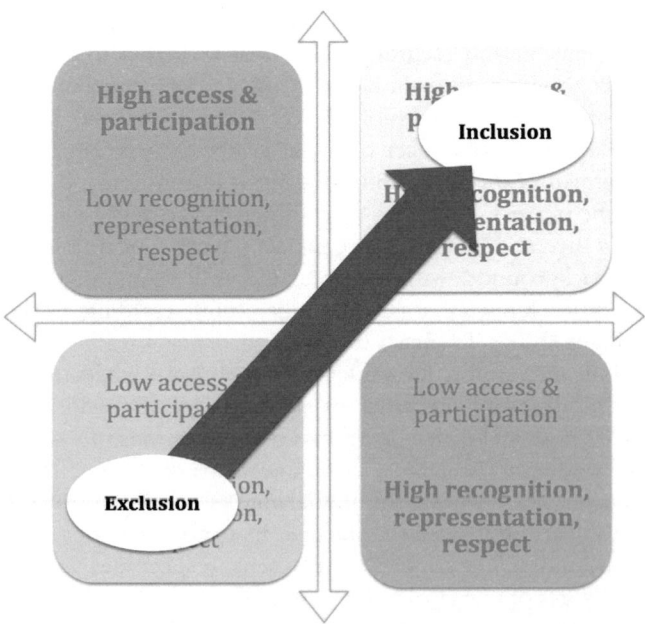

Figure 7.2 Inclusion and exclusion in the model of redistributive and relational social justice

It is also worth stating that in working with a spectrum of weak to strong forms of social justice I do not mean to imply that weaker versions of inclusion are not important. Often they are fundamental. They are however, rarely sufficient. As discussed in Chapter Six, physical access to science museums did not mean that participants found these spaces or the science learning opportunities therein accessible. Rather these visits were experienced as oppressive because the practices embedded in those institutions reinforced structural inequalities and participants' exclusion. Instead, thinking about social justice along a spectrum helps to foreground multiple perspectives and the importance of both redistributive and relational social justice in thinking through exclusion and inclusion in everyday science learning (see Figure 7.2).

This dual approach to social justice was key for understanding participants exclusion from *and* non-participation in everyday science learning, as discussed previously (see Figure 7.2). In turn, it anchors how I understand inclusion in everyday science learning practices. While Bourdieu's conceptual tools are useful for understanding how exclusion happens, for me, they are less helpful for turning exclusion on its head. Instead, I draw on the ideas of infrastructure access, literacies and community acceptance, framed by a view of inclusion that is based on both redistributive and relational social justice (Dawson, 2014a, 2017; Grabill, 1998; Porter, 1998).

Three lenses for accessible, equitable practice

The second part of my theory of inclusion for everyday science learning involves reframing what we have learnt about exclusion and social justice into tools to think specifically about everyday science learning practices. The concepts of infrastructure access, literacies and community acceptance were originally developed by Porter (1998) and Grabill (1998) in relation to community learning projects, and it is their community-centric approach that I find so useful. These three concepts serve as lenses, or levels of analysis, for understanding what might change across the everyday science learning system. Each can be thought about in terms of weak and strong forms of inclusion. As such they provide powerful tools for thinking about what might change for everyday science learning to become inclusive.

James Porter (1998, p. 103, italics in the original) argued " there are three ways to think about access: in an infrastructure sense as technical *resources* (i.e. the machine itself), in an educational sense as a technical *literacy* (i.e. the skills and expertise necessary to use the resources), and in a social sense as *community acceptance*". Clearly, as described by Porter (1998), these three concepts are fairly specific to community computing and Internet access. However, in combination with the theories and data from the rest of this book, these lenses can help us to reconfigure everyday science learning. Thus in the framework discussed in the next section of this chapter, I develop the concepts of infrastructure access, literacies and community acceptance as lenses for thinking about accessible and equitable everyday science learning, building on ideas from social justice, social reproduction and critical pedagogy (Bourdieu, 1984; Bourdieu & Passeron, 1990; Delpit, 2006; Fraser, 2003; Freire, 1998; hooks, 1994; Young, 1990, 2000).

An access and equity framework for everyday science learning

The framework I describe in this section represents my attempt to understand inclusion across the different fields of everyday science learning, following a theoretically and empirically informed approach. The framework operationalises the theory of inclusion for everyday science learning that I have discussed already. The framework comprises three lenses (see Figure 7.3). These three lenses (infrastructure access, literacies and community acceptance) overlap, highlighting the multifaceted nature of inclusion issues and the transformative potential of addressing multiple issues together.

In this section I discuss how each of the three lenses can be understood from a spectrum from weak to strong inclusion, illustrated with international examples

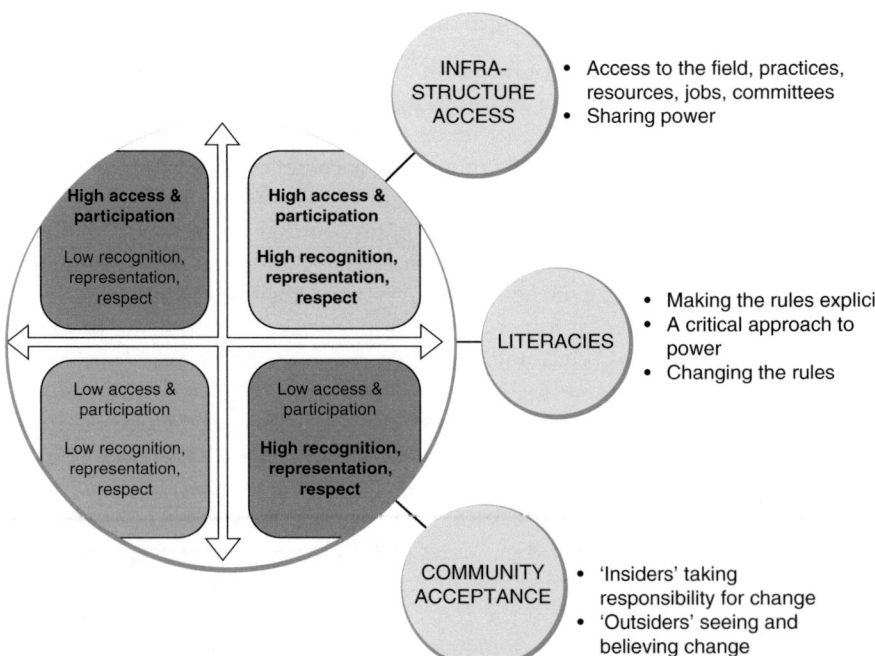

Figure 7.3 Three lenses for thinking about inclusion

of everyday science learning practice. Each lens can be understood in multiple ways (from weak to strong forms of inclusion) and, since they overlap, they build on one another. It is not enough therefore to think only about one or two of these lenses; inclusion requires that we think about all three. As a result, the framework can be used to think about a variety of approaches to inclusion in everyday science learning, as well as to understand which may be more likely to work and why some fail.

Understanding and transforming infrastructure access

I start with the concept of infrastructure access here because, as Porter (1998) has argued, infrastructure access plays a fundamental role in whether practices are inclusive. Without addressing inaccessibility, changes to everyday science practices that focus on literacies or community acceptance will make little difference. Infrastructure access therefore forms the necessary precondition for inclusion, whether in terms of practices across fields, within fields or individual activities. In other words, we have to ask the basic question: can people access the spaces, practices and resources involved in everyday science learning, whether online, on television, in politics, in science festivals, after-school clubs or in museums?

I read the concept of infrastructure access through Bourdieu's (1998; Bourdieu & Johnson, 1993) work on field. From this perspective infrastructure access can be understood not only as access to a field but also access to the practices and forms of valued capital within that field. I also draw on the dualist approach to social justice outlined previously to see issues of infrastructure access arranged on a spectrum from weak to strong versions of inclusion. The foundational part of infrastructure access can therefore be understood in terms of equality and redistributive social justice, as opening a restricted field or practice up to more people (Rawls, 1971).

As discussed in Chapter Three, everyday science learning operates as a series of restricted fields of production, excluding people in ways that, for these participants, were particularly marked by 'race'/ethnicity and class. As a result, everyday science learning can, I suggest, be best understood as only partially public and, for many people, far from 'everyday'. Thus, there is a significant redistributive aspect to infrastructure access in the context of everyday science learning. At a basic level, these fields, practices and the resources they contain need to be more available to more people if they are to be described as for 'the public'.

Infrastructure access for everyday science learning also requires relational social justice to be taken into account. As I discussed in Chapters Three, Five and Six the infrastructure of everyday science learning was inaccessible for participants by virtue of costs, geography, information, language and literacies and marketing as well as practices of representation rooted in racism, class discrimination and, for some participants, sexism. This is not a short list. The research carried out for this book shows how intertwined issues of racism and class discrimination were for participants' access to everyday science learning practices and how this played out in infrastructure access. For instance, entry costs, the price of food in a museum café, the price of toys in a science centre shop, the cost of transport across the city to get to a particular location were all examples of how economic capital structured the fields of everyday science learning. At the same time, people from racialised groups were either not represented (in languages, images, stories, texts, the kinds of foods available and in terms of the imagined and real bodies involved in everyday science learning practices) or portrayed as 'objects' rather than 'subjects', with little or no space made for their knowledges, practices or values. Issues of representation are greatly contested, and rightly so (Hall, 2013; Taylor, 1994). But addressing representation from a social justice perspective – and all the issues that it raises – seems more fruitful to me than falling back on out-dated, conservative and exclusive tropes of representation.

As Chapters Five and Six showed, even once physical infrastructure access was addressed in the form of physical entry to a museum space, practices of representation, cultural imperialism, powerlessness and practices about whose knowledge, practices or bodies mattered continued to exclude participants. As Chapter Six shows all too clearly, we need to think of infrastructure access in terms of equity as well as equality. Including a broader, more diverse public into unreconstructed forms of everyday science learning practice is, as discussed in Chapter Two, about assimilation, not about equality or equity. Instead, we will be better able to create meaningfully inclusive everyday science learning practices if we disrupt and transform exclusive practices. While I appreciate this kind of commitment requires significantly more resource, carrying out this research has convinced me that this dualist approach to inclusion is more likely to create inclusive everyday science learning practices than any 'quick fix' or attempt to address 'low hanging fruit' will.

To transform everyday science learning we need to see inclusive infrastructure access as determined by both equality and equity. Thus, activities that address only redistributive or relational social justice can be understood as weaker forms of inclusion than activities that address both. For example, a golden ticket scheme that waives entry costs to a museum without changing any of the practices inside addresses redistribution without addressing representation. As such it can be seen as a weak form of inclusion. A stronger form of inclusion would address issues of both literal, physical access (redistributive social justice) *and* the extent to which people have power and resources to shape a field or practice to suit their needs and value their assets (relational social justice).

To understand how a strong model of inclusion (one that combines relational and redistributive approaches) could work it is helpful to think not only about how users, audiences or visitors access a field, but how they could be involved in producing that field (Arnstein, 1969; Bienkowski, 2016; Bonney et al., 2009; Lynch, 2011; Powell & Colin, 2008; Simon, 2010). Power sharing can be considered a key part of improving infrastructure access, thus people are included not only as users, audiences or visitors, but as producers and decision makers. Porter (1998), for instance, argues that access to decision-making processes is a crucial element of infrastructure access, not least decisions about development and design processes. Some maker or hacker spaces have, for instance, worked to create spaces that prioritise the comfort, practices, knowledges and values of people from racialised groups and/or women (Maalsen & Perng, 2016; O'Sullivan, 2018; Toupin, 2014). These spaces developed in response to the criticism that maker or hacker spaces were spaces of white, male privilege, open to all in name only (Fox, Ulgado, & Rosner, 2015; Willett, 2016). Maker or hacker spaces that have prioritised equity and equality have sought to explicitly disrupt infrastructure access (as well as transforming literacies and community acceptance), by creating welcoming, safe spaces with, for instance, childcare facilities and different opening hours and have worked in partnership with specific communities to do so.

While the idea of being on a science festival advisory board was a long way away from participants' experiences (or wishes!), opening up who produces and is represented by everyday science learning across culture, education and politics is crucial for transforming practice (Das & Lowe, 2017; Lynch & Alberti, 2010; Puwar, 2001; Saha, 2018; Taylor & O'Brien, 2017). To explore this in everyday science learning would including looking at hiring practices for staff, volunteers, board members, the membership of institutional committees and the roles people hold within them, as well as how potential users, audiences or visitors are involved in content, programme, activity or exhibit development processes. Through processes such as these, issues involved in infrastructure access, literacies and community acceptance can be explored *with* excluded or non-participating groups by working in partnership, rather than attempting to second-guess their assets and needs.

Change to infrastructure access requires the redistribution of access to a field or practice as well as reshaping those fields and practices.

Disrupting everyday science learning literacies

The concept of literacies highlights the multiple, often hidden literacies required to be able to participate in everyday science learning (Dawson, 2014a). For instance, in monolingual after-school science clubs you may need to know the actual language used, to have a degree of scientific literacy as well as practical 'know-how' (such as how to use a specific tool) in order to be able to learn science. To think about equity and inclusion through the lens of literacies we have to ask what does it take to be able to navigate, use and benefit from the resources of the fields involved in everyday science learning?

As before, I understand literacies in terms of Bourdieu's work (1977, 1991; Bourdieu & Passeron, 1990); not only do people need to access a field (infrastructure access), but they need to have the skills to be able to play the games of that field successfully (a combination of habitus and capital). Literacies, from this perspective, can be considered in terms of cultural capital. These are forms of knowledge that support someone to achieve what they want in a field, whether that is finding the toilets in a museum or knowing how to behave in a town hall meeting about local pollution in order to make their voices heard. Literacies can also be usefully understood from a dualist social justice perspective. For instance, a weak reading of this concept focuses on surfacing the literacies, or forms of cultural capital, that facilitate access to everyday science learning, and supporting participants to develop the literacies they need (in other words, to change themselves). A strong reading of literacies would also focus on people's needs to have their practices, knowledges, forms of capital and values recognised and respected (Brown, 2006; Delpit, 1988, 2006; Gonzalez et al., 2013). In other words, a strong approach to inclusion involves not only supporting people to understand the rules of the game, but changing both the rules and the game itself. This approach can be thought of in terms of critical pedagogy. That is, applying a critical, strong approach to inclusion and the concept of literacies helps up to think about the roles power and structural inequalities play in whose literacies are recognised, valued and represented (Delpit, 1988, 2006; Freire, 1998; Freire & Freire, 1992; hooks, 1994).

Disrupting and transforming the way literacies are involved in everyday science learning is vital to developing inclusive, equitable practices. As discussed in Chapter Three, participants simply did not know about the range of everyday science learning practices that existed across their city or in their neighbourhoods. Not only were these practices largely invisible to participants, but they struggled to see their relevance or how they could benefit from them. Furthermore, as discussed in Chapter Six, once involved, participants found they needed a whole suite of literacies to navigate the museums and the science centre. They needed to speak and read English fluently, they needed scientific literacy (both in terms of actual language and the concepts involved), they needed museum literacy to know how to use exhibits, but they also needed knowledge about history, geography and other subjects that exhibits assumed knowledge of. Because the everyday science learning practices participants experienced were designed around assumptions about users' literacies, as I argued in Chapters Five and Six, participants had to do a significant amount of work to navigate these practices, to use the resources therein and, notably, to put themselves into a situation where they felt Othered. This research suggests therefore that attempts at inclusive practice could start by trying to disrupt some of these assumptions embedded in everyday science learning practices.

Importantly, taking a strong approach to developing inclusive everyday science learning helps us to think beyond the 'literal' literacies people need to successfully navigate a science club, laugh at a YouTube video about science or make themselves heard in a political consultation about science. While Porter's (1998) use of the term literacy as a core component of inclusive practice is valuable, it is crucial to stay alert to the implications of describing people as illiterate and the relationships between supposed literacy deficits and power. Building on work from critical pedagogy, we should think about whose selves, knowledges, languages and ways of being are recognised, represented and welcomed in everyday science learning practices and how these might be resisted, opened up or transformed (Delpit, 1988; Freire, 1998; hooks, 1994)? This is also the place to ask questions of content, especially science-related content. Science has a colonial, racist, sexist, homophobic, heteronormative, classed and ableist legacy, such that questions of representation must go beyond the faces shown in marketing leaflets to the very question of how knowledges are made and remade (Haraway, 1997, 1998; Longino, 1990; Medin & Bang, 2014; Schiebinger, 2007). This stronger form of inclusion has implications for changes in practice, institutions, policies and science, rather than only changing participants.

The Austrian Knowledge Rooms project provides a useful example of working towards equitable practice in a community youth science club setting. Organised by the Austrian Science Centre Network, the pop-up Knowledge Rooms used empty shop fronts in Vienna based in disadvantaged neighbourhoods as safe spaces for youth to engage with science (Austrian Science Centre Network, 2017; Streicher, Unterleitner, & Schulze, 2014). The rules of each Knowledge Room were created with participating youth and displayed on the walls of the space to help address how disoriented some youth felt in a science learning space. This work supports the inclusion of more people into everyday science learning by providing the tools to better navigate and thereby benefit from such practices (a weak form of inclusion, focused around a redistributive approach). It also starts to transform everyday science learning practices by working directly with youth and their families in partnership to transform the space, content and mediation practices into activities that are meaningful, relevant and valued by the young people (a stronger form of inclusion, combining redistributive and relational approaches to social justice).

Another interesting example can be seen in the *Sci-Girls* television show developed in the United States (PBS, 2017). This multiplatform programme aims specifically at creating female friendly science content, representing girls and women from a range of ethnic backgrounds exploring science on television and through educational outreach workshops. The project takes a strong approach to inclusion and literacies through the development of Spanish language episodes as well as addressing critical literacies through content that reflects on power in science (Knight-Williams et al., 2016). This approach involves representing and valuing a more pluralistic view of what science is and how people are involved in science on television (cf. Paulsen, 2013).

Taking a critical approach to literacies opens the door for reflecting on the assumptions, norms, expectations and power practices embedded within everyday science learning (Freire, 1998; hooks, 1994). Certain forms of knowledge and certain social groups are privileged in our societies. Without acknowledging how power operates in such systems it is hard to "offer strategies for social transformation" (Yosso, 2005, p. 71). Making the rules of the game more transparent and accessible, including the explicit and implicit literacies and codes needed, is one way of opening up everyday science learning to more people (Archer, Dawson, DeWitt, Seakins, & Wong, 2015; Brice Heath, 2007; Delpit, 1988, 2006). A more ambitious, stronger form of inclusion would be to disrupt the field such that the rules of the game were not only explicit to everyone, but changed such that people from minoritised backgrounds were no longer required to shoulder an additional workload whenever they want to participate in everyday science learning.

A critical approach to inclusion in everyday science learning has three steps in terms of the literacies lens: firstly, recognising that multiple forms of literacy are needed for participating successfully; secondly, acknowledging and making explicit the role of power and the established rules of the game; and thirdly, transforming everyday science learning practices to include the knowledges, skills and practices of minoritised groups.

Reconfiguring community acceptance

The final lens in the framework is the concept of community acceptance. For Porter (1998) community acceptance is about the people we might think of as insiders to a field or practice accepting newcomers (or people we might think of as outsiders) into their practices. Let's imagine a youth-led citizen science project as an example. Community acceptance, for Porter, is about youth already involved in the project welcoming newcomers, sharing resources, embracing new perspectives and restructuring infrastructure access and literacy practices as necessary to support newcomers.

As with the concepts of infrastructure access and literacies, I find it useful to think about community acceptance through Bourdieu's (Bourdieu, 1977; Bourdieu & Passeron, 1990) work on social reproduction and dualist approaches to social justice (Fraser, 2003). Combining these approaches with Porter's (1998) work can help us to understand community acceptance in a dual sense: first, to think about how existing stakeholders involved in everyday science learning welcome new participants and change their practices to do so (Dawson, 2014a; Porter, 1998); Second, to recognise and value the views, experiences and expectations of minoritised groups about whether everyday science learning opportunities seem relevant, useful and worth pursuing (Bourdieu, 1977; Bourdieu & Wacquant, 1992; Dawson, 2018; Fraser, 2003).

Importantly, this makes space for participants to reject as well as be excluded from everyday science learning, as discussed in Chapters Five and Six. Following a dualist approach to social justice as mentioned previously, considering people both inside and outside everyday science learning constitutes a stronger version of inclusion through the community acceptance lens than focusing only on one or the other. To draw on Bourdieu's (1977, 1991; Bourdieu & Passeron, 1990) work then, community acceptance can be understood as about habitus, symbolic violence and the extent to which habitus might shift, such that new ideas and practices could be created and embraced by those inside and outside a given field.

A key theme in this book has been about how participants' experiences affected their habitus and disposed them against rather than towards everyday science learning practices across a number of fields. With the exception of television and online activities, as discussed in Chapters Three and Five, participants, by and large did not participate in everyday science learning activities and had little interest in doing so. Furthermore, as discussed in Chapter Four, without disliking science, participants were disposed against it and shared experiences of marginalisation in relation to science. Such deeply held views and the experiences that they grow from are unlikely to change overnight or without significant effort.

Clearly it is important that producers (staff, volunteers, policy makers, board members) and included audiences welcome people who are currently excluded from everyday science learning for change to happen. These 'insiders' must value and invest in the processes of disruption, transformation and change necessary to develop inclusive practices. That minoritised groups continue to shoulder the burden of labour for 'fitting in' to unreconstructed everyday science learning practices is clearly unjust. Change is necessary because, as Young has argued, "bringing about justice where there is exploitation requires reorganisation of institutions and practices in decision-making, alteration of the division of labour, and similar measures of institutional, structural, and cultural change" (1990, p. 53). But embracing change is not always easy. The notion of community acceptance as described by Porter (1998) and Grabill (1998) requires those involved in everyday science learning practices to take responsibility for exclusive practices and to address structural inequalities, not least problems with infrastructure and literacies. Such changes will come with costs, not least the resources needed to significantly reorganise current practices and the emotional costs of letting go of power, privilege and dearly held ways of working.

From the data discussed throughout this book it is clear that community acceptance cuts both ways. What kinds of change would be needed for Fatimata, Abdou, Sarasa, Fatima or Alejandro to re-evaluate their taken-for-granted disposition again everyday science learning? Put another way, community acceptance also requires people who view everyday science learning as irrelevant to their lives to shift their dispositions towards, rather than away from, everyday science learning. This is a difficult task because, as discussed in Chapters Three, Four, Five and Six, everyday science learning did not represent a compelling, appealing, relevant or valuable set of fields or practices as they were currently configured for participants. As Chapters Three through Six show, participants' dispositions against everyday science learning had deep roots in socio-historic, political histories of oppression and structural inequalities, none of which are easy to disrupt and transform. We need to think carefully about what would be involved in developing anti-racist, anti-sexist, anti-classist everyday science learning practices. How can we address these three issues together as well as issues of ableism, ageism, heteronormativity and homophobia and the many other forms of oppression reflected in everyday science learning?

Here, as before, working in partnership may represent a significant opportunity for usefully reshaping practice to be more inclusive. A useful example comes from community partnership work carried out between local government, the local environmental agency, a local community and a science centres (Parque Explora) in Medellin, Colombia. Working together, these groups built long-term, mutually beneficial relationships to address housing issues and environmental problems and to leverage educational assets and needs (Aguirre, 2014). Power was devolved from the municipal organisations to community leaders, their networks and trained community members who worked as facilitators. This led to a significant amount of community participation in terms of the shape of the project and who held decision-making power. In this example a strong approach to inclusion and community acceptance was enrolled into the participatory practices that guided how the project worked. Similarly in the UK, museums and archives involved in the Our Museum project worked to restructure their organisations to embed participatory practices across all their work. They transformed decision-making processes, opened-up programme and exhibition design, reworked hiring practices and, ultimately, changed the profile of their participants (Bienkowski, 2016; Lynch, 2011).

A critical approach to understanding inclusion and community acceptance for everyday science learning needs to be understood as multidirectional. It must be relevant to stakeholders who already produce and use everyday science learning practices and those who do not already participate.

Putting it all together

These three lenses – infrastructure access, literacies and community acceptance – can be understood in terms of social justice to help us think about what would be involved in inclusive everyday science learning. I have come to see each lens as working with the next in multiple, mutually enhancing ways. Since each lens can be understood on a continuum from weak to strong forms of inclusion, the framework can be used to think about how different inclusion practices complement or undermine one another.

For instance, in Figure 7.4, which follows, the top right box (numbered 1) represents what I described earlier as the sweet spot for inclusive practice. Practices here consider both relational and redistributive aspects of social justice across all three lenses. An example could be a citizen science group where people from minoritised backgrounds develop long-term, citizen science projects that are relevant to their lives. These projects could draw on their assets and provide mechanisms to support their goals, both within and beyond science. Imagine if this process involved not only recognising and representing people's values, practices and knowledges, but valued and celebrated them. Notably, for people in dominant social groups, the experiences described in box 1 can be thought of as standard. In contrast, in Figure 7.4, the bottom left box (numbered 4) represents the more exclusive, inequitable forms of practice, which the rest of this book provides examples of.

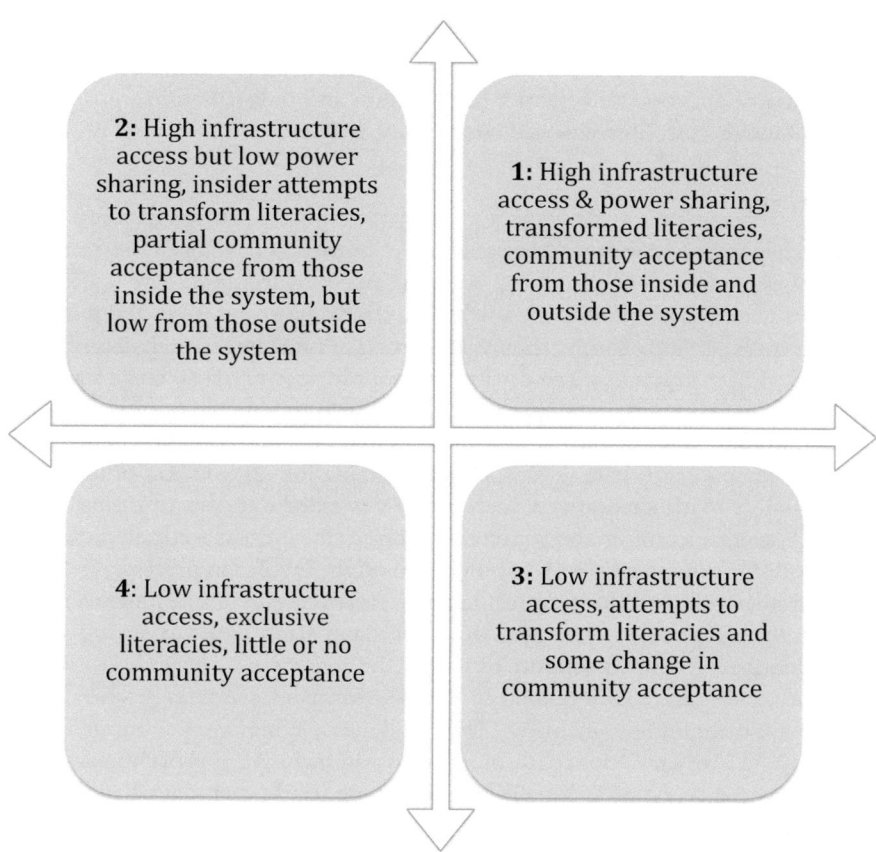

Figure 7.4 An access and equity framework for everyday science learning

The other two boxes (numbered 2 and 3) represent interesting ways to think about partial attempts at inclusive practice. In box 2, for instance, we could think about attempts to generate more viewers, users, audiences or visitors but without challenging or changing who and what is represented, such as the golden ticket practices discussed in Chapter Two. Similarly in box 3, we could think about attempts to change whose bodies, knowledges, practices and stories are represented in a museum setting where entry fee structures and marketing practices remain unchanged and, as a result, both forms of community acceptance remain limited (Lynch & Alberti, 2010). The framework therefore provides a thinking tool to strategically plan and evaluate practice, not least how and why certain practices may not be supporting inclusion as they might be intended to.

Nancy Fraser argued that for a dualist view of social justice "the goal, in its most plausible form, is a difference-friendly world, where assimilation to majority or dominant cultural norms is no longer the price of equal respect" (2003, p. 7).

In other words, developing inclusive, equitable everyday science learning practices requires more than providing existing activities to different kinds of people (assimilation and redistribution), but taking difference into account and developing difference-appropriate activities (recognition and redistribution) in terms of infrastructure access, literacies and community acceptance.

Summary

In this chapter I have argued that there is potential to develop more inclusive forms of everyday science learning. Building on the admittedly few but salient moments where participants were able to use their own knowledges and practices to connect with science during their visits carried out for the research described in this book, I have argued we can disrupt and transform everyday science learning, *if we choose to*. These changes, as I have argued throughout the book, are necessary and urgent. To continue with business as usual is to be complicit in practices that uphold and exacerbate racism, class discrimination, sexism and other forms of oppression. With change as a goal, I have described a theory of inclusion for everyday science learning and outlined a empirically and theoretically informed framework for thinking about what inclusion might involve in practice.

The framework uses three critical lenses – infrastructure access, literacies and community acceptance – arranged on a spectrum from weak to strong forms of inclusion. As a tool to support thinking, the framework can be used to plan and to evaluate practices intended to develop forms of everyday science learning that are meaningfully inclusive. Thus while people may well continue not to participate in everyday science learning, taking an inclusive approach means that practices where everyone is equally able to choose to use everyday science learning (or not), to have meaningful, relevant and respectful experiences where their own communities, knowledges and practices are welcome should be our goal.

Note

1 Ok, maybe it's just me asking this question! But the truth is I ask this question a lot and it's something I struggle with.

References

Aguirre, C. (2014). Science centers. Which roles can they play to participate in a city social reconstruction? *Journal of Science Communication, 13*(2), 1–12.
Ahmed, S. (2000). *Strange encounters: Embodied others in post coloniality.* London: Routledge.
Ahmed, S. (2012). *On being included: Racism and diversity in institutional life.* Durham and London: Duke University Press.
Aikenhead, G. (2002). Science communication with the public: A cross-cultural event. In W.-M. Roth & J. Désautels (Eds.), *Science education as/for sociopolitical action* (pp. 151–166). New York: Peter Lang Publishing.

Archer, L., Dawson, E., DeWitt, J., Seakins, A., & Wong, B. (2015). "Science capital": A conceptual, methodological, and empirical argument for extending bourdieusian notions of capital beyond the arts. *Journal of Research in Science Teaching*, 52, 922–948. doi.org/10.1002/tea.21227

Archer, L., Dawson, E., Seakins, A., & Wong, B. (2016). Disorientating, fun or meaningful? Disadvantaged families' experiences of a science museum visit. *Cultural Studies of Science Education*, 1–23. doi:10.1007/s11422-015-9667-7

Arnstein, S. R. (1969). A ladder of citizen participation. *Journal of the American Institute of Planners*, 35(4), 215–224.

Austrian Science Centre Network. (2017). Retrieved from www.science-center-net. at/index.php?id=621, accessed 14.03.2017

Bhabha, H. K. (1994). *The location of culture*. Abingdon and New York: Routledge.

Bienkowski, P. (2016). *No longer us and them: How to change into a participatory museum and gallery*. London: Paul Hamlyn Foundation.

Bonney, R., Cooper, C. B., Dickinson, J., Kelling, S., Phillips, T., Rosenberg, K. V., & Shirk, J. (2009). Citizen science: A developing tool for expanding science knowledge and scientific literacy. *BioScience*, 59(11), 977–984. doi:10.1525/bio.2009.59.11.9

Bourdieu, P. (1977). *Outline of a theory of practice* (R. Nice, Trans.). Cambridge: Cambridge University Press.

Bourdieu, P. (1984). *Distinction: A social critique of the judgement of taste* (R. Nice, Trans.). London: Routledge.

Bourdieu, P. (1991). *Language and symbolic power* (G. Raymond & M. Adamson, Trans.). Cambridge and Malden, MA: Polity Press.

Bourdieu, P. (1998). *Practical reason*. Cambridge: Polity Press.

Bourdieu, P., & Johnson, R. (1993). *The field of cultural production: Essays on art and literature*. Cambridge: Polity Press.

Bourdieu, P., & Passeron, J.-C. (1990). *Reproduction in education, society and culture* (R. Nice, Trans., 2nd ed.). London, Newbury Park, CA and New Delhi: Sage.

Bourdieu, P., & Wacquant, L. (1992). *An invitation to reflexive sociology*. Chicago: University of Chicago Press.

Brice Heath, S. (2007). *Diverse learning and learner diversity in "Informal" science learning environments*. Palo Alto National Research Council Board on Science Education.

Brown, B. A. (2006). "It isn't no slang that can be said about this stuff": Language, identity, and appropriating science discourse. *Journal of Research in Science Teaching*, 43, 96–126. doi.org/10.1002/tea.20096

Dancstep, T. (February 2018). London: Personal communication.

Das, S., & Lowe, M. (2017). *Nature read in Black and White, or how to stop being racist and develop meaningful natural history curation*. Paper presented at the Natural Sciences Collections Assocation (NatSCA) conference, University Museum of Zoology, Cambridge, UK.

Dawson, E. (2014a). Equity in informal science education: Developing an access and equity framework for science museums and science centres. *Studies in Science Education*, 50(2), 209–247. doi.org/10.1080/03057267.2014.957558

Dawson, E. (2014b). "Not designed for us": How science museums and science centers socially exclude low-income, minority ethnic groups. *Science Education*, 98(6), 981–1008. doi.org/10.1002/sce.21133

Dawson, E. (2017). Social justice and out-of-school science learning: Exploring equity in science television, science clubs and maker spaces. *Science Education*, *101*(4), 539–547. doi:10.1002/sce.21288

Dawson, E. (2018). Reimagining publics and (non)participation: Exploring exclusion from science communication through the experiences of low-income, minority ethnic groups. *Public Understanding of Science*, 27(7), 772–786. doi:10.1177/0963662517750072

Delpit, L. (1988). The silenced dialogue: Power and pedagogy in educating other people's children. *Harvard Educational Review*, *58*(3), 280–299.

Delpit, L. (2006). *Other people's children: Cultural conflict in the classroom*. New York: New Press.

Eisenhart, M. (2001). Educational ethnography past, present, and future: Ideas to think with. *Educational Researcher*, *30*(8), 16–27.

Fox, S., Ulgado, R. R., & Rosner, D. (2015). *Hacking culture, not devices: Access and recognition in feminist hackerspaces*. Paper presented at the 18th ACM conference on Computer Supported Cooperative Work & Social Computing, Vancouver, Canada, BC.

Fraser, N. (2003). Social justice in the age of identity politics: Redistribution, recognition, and participation. In N. Fraser & A. Honneth (Eds.), *Redistribution or recognition? A political-philosophical exchange* (pp. 7–109). London and New York: Verso.

Fraser, N., & Honneth, A. (2003). *Redistribution or recognition? A political-philosophical exchange*. London and New York: Verso.

Freire, P. (1998). *Pedagogy of freedom: Ethics, democracy, and civic courage*. Lanham, Boulder, New York and Oxford: Rowman & Littlefield Publishers.

Freire, P., & Freire, A. M. A. (1992). *Pedagogy of hope: Reliving pedagogy of the oppressed*. New York: Continuum.

Gonzalez, N., Moll, L. C., & Amanti, C. (2013). *Funds of knowledge: Theorizing practices in households, communities, and classrooms*. Mahwah, NJ: Taylor & Francis.

Grabill, J. T. (1998). Utopic visions, the technopoor, and public access: Writing technologies in a community literacy program. *Computers and Composition*, *15*(3), 297–315.

Hall, S. (2013). The spectacle of the "Other". In S. Hall, J. Evans, & S. Nixon (Eds.), *Representation* (2nd ed., pp. 215–287). London and New Delhi: Sage.

Hammonds, E. (2009). *The nature of difference: Sciences of race in the United States from Jefferson to genomics*. Cambridge, MA: MIT Press.

Haraway, D. (1997). *Modest_Witness@Second_Millennium.FemaleMan©Meets_Onco Mouse™: Feminism and technoscience*. New York: Routledge.

Haraway, D. (1998). Deanimations: Maps and portraits of life itself. In C. A. Jones & P. Galison (Eds.), *Picturing science, producing art* (pp. 181–207). London: Routledge.

Harding, S. (2008). *Sciences from below: Feminisms, postcolonialisms, and modernities*. Durham: Duke University Press.

hooks, b. (1994). *Teaching to transgress: Education as the practice of freedom*. London and New York: Routledge.

Johnson, A. (2017). Getting comfortable to feel at home: Clothing practices of black muslim women in Britain. *Gender, Place & Culture*, 24(2), 274–287. doi:10.1080/0966369X.2017.1298571

Knight-Williams, V., Williams, D., Teel, R., Williams, E., Hernandez, M., Simmons, G., . . . Rahbari, S. (2016). *SciGirls season four: Front-end evaluation report.* Retrieved from www.scigirlsconnect.org/wp-content/uploads/2016/06/Latina-SciGirls-Front-End-Evaluation-Report-FINAL-6.22.2017.pdf

Longino, H. E. (1990). *Science as social knowledge: Values and objectivity in scientific inquiry.* Princeton, NJ: Princeton University Press.

Lorde, A. (1984). *Sister outsider.* Berkeley: Crossing Press.

Lynch, B. (2011). *Whose cake is it anyway? A collaborative investigation into engagement and participation in 12 museums and galleries in the UK.* London: Paul Hamlyn Foundation.

Lynch, B., & Alberti, S. J. M. M. (2010). Legacies of prejudice: Racism, co-production and radical trust in the museum. *Museum Management and Curatorship, 25*(1), 13–35. doi.org/10.1080/09647770903529061

Maalsen, S., & Perng, S.-Y. (2016). Crafting code: Gender, coding and spatial hybridity in the events of Pyladies Dublin. *The programmable city working paper (19).*

Mai, T., & Ash, D. (2012). Tracing our methodological steps. In D. Ash, J. Rahm, & L. Melber (Eds.), *Putting theory into practice: Tools for research in informal settings.* Rotterdam: Sense Publishing.

Medin, D. L., & Bang, M. (2014). *Who's asking? Native science, western science, and science education.* Cambridge, MA and London: MIT Press.

Miles, A., & Gibson, L. (2016). Everyday participation and cultural value. *Cultural Trends, 25*(3), 151–157. doi.org/10.1080/09548963.2016.1204043

Museum Detox. (2018). *Museum Detox: A neworking group for BAME professionals in museums and heritage.* Retrieved from http://museumdetox.com/#team-5

O'Sullivan, E. (2018). Excellence in the maker movement. *Journal of Peer Production, 3*(12), 46–50.

Paulsen, C. A. (2013). Implementing out-of-school time STEM resources: Best pratices from public television. *Afterschool Matters,* (17), 27–35.

PBS. (2017). *SciGirls.* Retrieved from www.pbs.org/parents/scigirls

Porter, J. E. (1998). *Rhetorical ethics and internetworked writing.* Greenwich, CT and London: Ablex Pub. Corp.

Powell, M. C., & Colin, M. (2008). Meaningful citizen engagement in science and technology: What would it really take? *Science Communication, 30*(1), 126–136. doi:10.1177/1075547008320520

Puwar, N. (2001). The racialised somatic norm and the senior civil service. *Sociology, 35*(3), 651–670. doi:10.1017/S0038038501000335

Puwar, N. (2004). *Space invaders: Race, gender and bodies out of place.* Oxford and New York: Berg.

Rahm, J. (2010). *Science in the making at the margin: A multisited ethnography of learning and becoming in an afterschool program, a garden and a math and science upward bound program.* Rotterdam, Boston and Taipei: Sense Publishers.

Rawls, J. (1971). *A theory of justice.* Cambridge, MA: Harvard University Press.

Roth, W.-M. (2008). Bricolage, métissage, hybridity, heterogeneity, diaspora: Concepts for thinking science education in the 21st century. *Cultural Studies of Science Education, 3*(4), 891–916. doi.org/10.1007/s11422-008-9113-1

Saha, A. (2018). *Race and the cultural industries.* Cambridge: Polity Press.

Sandell, R., Dodd, J., & Garland-Thomson, R. (2010). *Re-presenting disability: Activism and agency in the museum.* Abingdon and New York: Routledge.

Schiebinger, L. (2007). Getting more women into science: Knowledge issues. *Harvard Journal of Law and Gender, 30*, 365–378.

Simon, N. (2010). *The participatory museum.* Santa Cruz: Museum 2.0.

Spivak, G. C. (1999). *A critique of postcolonial reason: Toward a history of the vanishing present.* Cambridge, MA and London: Harvard University Press.

Streicher, B., Unterleitner, K., & Schulze, H. (2014). Knowledge rooms: Science communication in local, welcoming spaces to foster social inclusion. *Journal of Science Communication, 13*(2), 1–5.

TallBear, K. (2013). Genomic articulations of indigeneity. *Social Studies of Science, 43*(4), 509–533. doi:10.1177/0306312713483893

Taylor, C. (1994). *Multiculturalism: Examining the politics of recognition* (A. Gutmann, Ed.). Princeton, NJ: Princeton University Press.

Taylor, M., & O'Brien, D. (2017). "Culture is a meritocracy": Why creative workers' attitudes may reinforce social inequality. *Sociological Research Online, 22*(4), 27–47. doi:10.1177/1360780417726732

Toupin, S. (2014). Feminist hackerspaces: The synthesis of feminist and hacker cultures. *Journal of Peer Production,* (5). Retrieved from http://peerproduction.net/issues/issue-5-shared-machine-shops/peer-reviewed-articles/feminist-hackerspaces-the-synthesis-of-feminist-and-hacker-cultures/

Trienekens, S. (2002). "Colourful" distinction: The role of ethnicity and ethnic orientation in cultural consumption. *Poetics, 30*(4), 281–298.

Willett, R. (2016). Making, makers, and makerspaces: A discourse analysis of professional journal articles and blog posts about makerspaces in public libraries. *The Library Quarterly, 86*(3), 313–329. doi:10.1086/686676

Yalowitz, S., Garibay, C., Renner, N., & Plaza, C. (2013). *Bilingual exhibit research initiative: Institutional and intergenerational experiences with bilingual exhibitions.* Washington, DC: Centre for the Advancement of Informal Science Learning.

Yosso, T. J. (2005). Whose culture has capital? A critical race theory discussion of community cultural wealth. *Race Ethnicity and Education, 8*(1), 69–91.

Young, I. M. (1988). Five faces of oppression. *The Philosophical Forum, 19*(4), 270–290.

Young, I. M. (1990). *Justice and the politics of difference.* Princeton, NJ: Princeton University Press.

Young, I. M. (2000). *Inclusion and democracy.* Oxford and New York: Oxford University Press.

Chapter 8

Afterword

This book has explored what happens when people from minoritised and, specifically, racialised groups encounter science through a variety of everyday science learning practices from a sociological perspective, while foregrounding issues of social justice. It is traditional for a final chapter to discuss things like the implications of research for practice, the limitations of the research and ideas for future research. Those of you reading this book in order will have noticed however, that the previous chapter focused at length on the implications of the research carried out of this book for understanding inclusion and equity, both in theory and in practice, along with a call to action. As a result, it is worth noting that this is not a traditional conclusions chapter. Instead I have titled it 'Afterword' and use it as a space to discuss some of the ideas that fall out of this research, in particular, questions about science content, racism and thinking beyond everyday science learning. This chapter finishes by taking a snapshot of equity in the contemporary everyday science learning landscape, where I discuss both my cynicism and greed about what change is possible.

To be clear, the key implication of the research I have discussed in this book is that everyday science learning is exclusive. This exclusion is embedded in the structure of the fields involved in everyday science learning and in their practices. Thus, as discussed in Chapter Seven, we need to do more than tweak staff training or marketing materials if we want everyday science learning to be truly public. We need to radically overhaul the systems, practices and ways of thinking that shape everyday science learning. Furthermore, we need to go beyond that to think about how inclusive everyday science learning practices can contribute more widely to disrupting and transforming social inequalities. I appreciate this may seem overwhelming, but I do not believe that minimising the need for change serves anyone well, least of all people from minoritised groups, like the participants who took part in the research I have discussed here.

On one hand, this is a book about everyday science learning practices – how they reproduce exclusive, oppressive practices of discrimination based on 'race'/ethnicity, class, gender and other facets of people's intersecting subjectivities. It is about how we can learn from experiences of exclusion to disrupt and transform everyday science learning such that the knowledges, practices and selves of people from racialised groups are recognised, represented and respected. On the other hand, this book can also be read as an example of how institutional racism and institutional class discrimination work together (not to mention how these are inflected in a kaleidoscopic manner with gender and the other facets of peoples' lives). In particular this book can be used as an example of how structural inequalities in the UK operate to protect and reify middle- and upper-class whiteness. In this second reading the context that frames the research (everyday science learning practices) provides a case study through which the reproduction of social disadvantages, not least the interplay of racism and class discrimination, can be unpacked. These two different perspectives provoke different ways to develop the work started here, as I discuss later.

Disrupting and transforming science

The research carried out for this book shows just how exclusive everyday science learning practices can be and the damage that exclusion does to people at the sharp end of these practices. As I have argued throughout this book, people who feel included in everyday science learning, whether as researchers, practitioners and producers, policy makers or included users, have a responsibility to centre social justice in their work. If we do not, we have to ask serious questions about whether we are happy to reproduce advantages for dominant groups at the expense of the minoritised. Two significant questions remain unanswered for me from the research reported in this book. First, how do we change what counts as science content in everyday science learning? And second, how can we think seriously about science practices above and beyond those involved in everyday science learning?

Content matters in everyday science learning. As discussed in Chapters Two, Four, Five and Six, science-related content was rarely welcoming for participants. Instead, science was represented in ways that reproduced tired tropes of being stale, pale and male (Lawler, 1996). As such I argued in this book that from a social justice perspective it is important to reconsider the science question, not least in terms of whose knowledges and practices count when it comes to thinking about science content in everyday science learning practices. I repeat this point here because one trend I see in attempts at equity work in everyday science learning is that people are more comfortable focusing on *how* science is communicated rather than the science content itself, as though science content is fixed and sacrosanct. As I discussed in Chapter Four, despite having a wealth of science-related knowledge and practices, even participants echoed extremely conservative views of what could and could not be called science.

Although I touched on this in Chapter Four, trying to understand how to usefully and meaningfully reconfigure what counts as science in everyday science learning seems to me to be a key area worthy of further work. Happily, there are practices and ideas that already exist that we might be able to use. Researchers working in science education on indigenous knowledge have long since debated how these forms of knowledge can be reconciled with science in ways that are neither appropriative nor patronising (Bang et al., 2014; Kim, Asghar, & Jordan, 2017). What if the science content of an everyday science learning project did not focus on knowledges that had their roots in the European 'enlightenment'? In practical terms, what could we learn from combining practices about the co-construction of scientific knowledge from citizen science practices with arts practices that seek to revalue how public art is created and understood (Ballard, Dixon, & Harris, 2017; Gibson, 2001). How might we use these practices and ideas to reimagine the epistemic practices of science? Although I do not have a clear sense of the answers, questions about how we work to open up the space of what counts as science seem crucial for developing meaningfully inclusive, equitable everyday science learning practices.

The question of science content is one that travels beyond everyday science learning and has implications for what we consider as the goals of social justice in relation to science. My second concern is about the relationships between developing inclusive, equitable everyday science learning practices and the broader worlds of science education and scientific careers. It is all very well creating everyday science learning practices that support, for instance, youth from minoritised backgrounds to feel comfortable using and producing science in a citizen science project. But we have to recognise that should those youth decide to pursue their science-related skills beyond such a project the environment they will meet will not necessarily be a friendly one. What is achieved if, for instance, a young Black woman from such a project goes on to study a PhD in physics only to experience the exclusive and damaging practices that we know take place in science in higher education (Gonsalves, Danielson, & Pettersson, 2016; Johnson, Brown, Carlone, & Cuevas, 2011; Ong, 2005)?

It seems to me we must consider disrupting and transforming science and science-related practices as well as everyday science learning. Indeed, as Londa Schiebinger (2007) has argued, a significant question remains about what kinds of science knowledges, skills and practices count and how they are counted. I remain convinced however that, as others have argued, from an epistemological perspective, a social justice perspective and a practice perspective, the most interesting, valuable, adaptive and ultimately useful forms of science will come from an inclusive scientific community (Longino, 1990; Medin & Bang, 2014).

Racism, knowledge and research

This book has been about what happens when people from racialised groups encounter science in the public sphere, through activities that I have described here as everyday science learning practices. In this book I have tried to tell a story about how structural inequalities – racism, class discrimination and sexism in particular – shape and are exacerbated by everyday science learning practices. Although I took an intersectional approach, as the story has unfolded I focused more and more on how the various intersecting structural inequalities participants experienced are rooted in racism and their position as racialised groups in the UK.

Ultimately therefore, this book is about racism. It is about the experiences of Latin American families and Sierra Leonean elders. It is about the stories of Fatima, Ibrahim and Mr Bhakta. It is about how institutional racism structures visits to science museums. It is about trying to change everyday science learning practices such that racism and associated practices of racialised class discrimination, racialised sexism and their intersections are recognised for what they are, so they might be addressed, disrupted and transformed.

As discussed in Chapter One, this book is based on qualitative, ethnographic research with five community groups in London, over a two-year period. It was exploratory research, designed to develop a better understanding of how people from racialised groups, living in relative poverty, experienced everyday science learning. Drawing on Mariana Ortega's (2006) work about whose experiences, practices and knowledges we attend to and how this affects knowledge making, this research was about providing a different point of view to that which dominates the literature. What this book offers therefore are the insights and experiences of people whose stories are not usually taken into account across the fields of everyday science learning. As Mitsuye Yamada (2015/1981, p. 69) wrote, quoting Cherríe Moraga "what each of us needs to do about what we don't know is to go look for it". This book is my attempt to listen and learn, as well as to share.

But in my experience not everyone finds it easy to think, listen or talk about exclusion, or racism in particular, in everyday science learning. Thus, specific but potentially undermining questions about generalisability[1] arise from time to time. Clearly, this research, rooted in qualitative and ethnographic methods, will not speak to everyone's experiences. That people across five different community groups, who never met one another, told stories that echoed each other's so closely is still valuable information to me. Thus, by being transparent about the methods used, involving participants in the analysis and presenting this work as clearly as I can, I hope to assure you the examples drawn on in this book are meaningful (please see the appendix for further details about research design and analysis).

In a similar vein, I am also frequently asked about how generalisable the research carried out for this book might be to white, working-class groups. This is a question all too often asked at conferences, to the extent that I now include my answer in the body of my presentations. That there are, yes, doubtless people from white working-class backgrounds who share aspects of participants' experiences of exclusion from everyday science learning. As to exactly how, who, why, where or when, I cannot say on the basis of this research. What's more, I am not only reluctant to speculate given the potential for damage, but because I have experienced this question too many times as a way to derail talk about racism. While I think it is of course important to think about exclusion as multiple and intersectional, there is no getting away from the fact that this study is about people from racialised groups. I have therefore found it more and more necessary to insist on 'race'/ethnicity and racism as key parts of this story, not least to prevent them being whitewashed away. Thus I have pushed to retain some specificity of language around racism and 'race'/ethnicity amidst words like 'exclusion' and 'social justice'. Perhaps I have still not pushed hard enough.

Beyond science: disrupting and transforming culture, education and politics

Read as a case study of how institutional practices across the different fields involved in everyday science learning – from the mass media, to museums, to schools, to political processes – are shaped by and in turn reproduce structural inequalities, this book has implications that travel beyond spaces where science is represented. For a start, aspects of the racial and classed discrimination participants experienced during the accompanied visits to the science museums and a science centre had nothing to do with content. Instead, racial profiling by security guards such that they were followed and repeatedly asked not to touch exhibits, or being asked to leave cafés before other visitors, are behaviours that are unrelated to learning science. As such, paying attention to practices of institutional racism, class discrimination, sexism and other forms of oppression across spaces that have nothing to do with science seems important.

Content, of various forms, does matter though and the research discussed in this book provides an example of how practices related to content – in this case to science – are marked by structural inequalities. Might this have implications further afield? If we take the field of art museums for instance, we can see from research by Cecelia Garibay (2017) and Catherine Hahn (2016) that issues of 'race'/ethnicity and racism play out in terms of the knowledges, practices and content on display, as well as in visitor patterns. If we turn to the museum field more broadly, research from Carol Dixon (2012, 2016), Viv Golding (2009) and the American Alliance of Museums (2018) show that questions of power, 'race'/ethnicity and structural inequalities continue to shape the museum landscape. And if we think too about research from Richard Sandell (2007) and Amy Levin (2010) we can see that similar struggles with representation, respect and recognition play out in museums in terms of (dis)ability, sexuality, gender, class and the intersecting subjectivities of people's lives. Thus, as the slogan I see increasingly frequently on T-shirts and Twitter puts it, 'museums are not neutral'. The research discussed in this book suggests the same is true for science in the mass media as well as public engagement with science practices in higher education and politics. Building on the parallels between international museum studies and the research carried out for this book suggests that the participants' experiences discussed here may have wider relevance beyond science-related settings.

As such, thinking beyond the science context of this book, we might think about how partial patterns of participation play out in television watching, enjoying sports or voting and what those patterns mean in terms of structural inequalities in our societies. What can we learn from how the ideas discussed in this book spread (or not) into research about who gets to produce content in the cultural, educational and political industries and the various influences on their agency (Dent, 2016; Puwar, 2001; Saha, 2018; Taylor & O'Brien, 2017)? What might thinking about publics in terms of structural inequalities mean for debates about public engagement and participatory democracy? How can concepts such as Fraser's (1990) micro-publics be developed in meaningful ways across these different contexts?

For scholars of social justice, being unable to participate in, benefit from or otherwise shape valued public practices, whatever they may be, constitutes a significant form of marginalisation and oppression (Fraser, 2003; Young, 1990). Ideas developed in this book about how inclusive, participatory and democratic models of various publics might be understood therefore apply beyond everyday science learning. If we understand publics as heterogeneous and active in global, multi-cultural societies, we can reimagine practices such that differences are valued rather than erased (Benhabib, 2002; Young, 1990, 2000). I make this point because as Puwar (2004) has argued, not everyone gets to be included in the public, yet ideas about what is public and who the public are lie at the heart of our societies. Thus it is helpful to keep in mind, as discussed in Chapter Two, that publics are brought into being in the light or shadow of specific practices, and that ideas from social justice can help to provide a framework for understanding what these practices of inclusion and exclusion mean.

If we consider any educational, cultural or political practices to be socially or personally valuable therefore, we must consider how exclusion operates and what equitable systems could look like. From this perspective, for example, an inclusive, empowering experience would be one that involved multiple voices, spaces and publics in equitable ways.

Taking the temperature of the water: a snapshot

I do not believe that inclusive, equitable everyday science learning practices are beyond our grasp. I appreciate how overwhelming the influence of structural inequalities can seem however, even when a person, team or organisation *is* committed to change. You can't just wish your way out of structural inequalities and institutional racism. Change takes work, making mistakes, admitting what we do not know, picking ourselves up and starting again. But it is possible. For instance, research in the US with Latinx communities found science museums were expected to be unfriendly places, with hard-to-understand exhibits (Garibay, 2009). When their languages *were* represented, however, people from Spanish speaking backgrounds in the US felt more valued by science museums, more comfortable during their visits and felt the museums were more culturally relevant (Yalowitz, Garibay, Renner, & Plaza, 2013). This is no small change.

The empirical and theoretical accounts of exclusion and inclusion in everyday science learning that I could not find at the start of this project are becoming more and more available (Feinstein, 2017; Garibay & Huerta Migus, 2014; Philip & Azevedo, 2017). Staying with museums as an example, issues of access, equity and inclusion seem to be garnering attention in ways that I desperately hope are meaningful. The work documented by Gretchen Jennings and Joanne Jones-Rizzi (2017) testifies to the deep commitment of particular practitioners and institutions to equity in science museums. Building on the "Race: are we so different" exhibition they developed with the American Anthropological Association, the Science Museum of Minnesota's anti-racist approach to everyday science learning places racial equity at the heart of all their activities, from exhibition content to staff recruitment (American Anthropological Association, 2018; Science Museum of Minnesota, 2018). Furthermore, inclusion and equity no longer seem to be single-issue concerns. It gladdens my heart to be able to reference projects that have addressed inclusion in museums from multiple and overlapping perspectives (see for instance, Achiam & Holmegaard, 2017; Dancstep & Sindorf, 2018; Garibay, Lannes, & González, 2018; Sandell, Dodd, & Garland-Thomson, 2010). And the stories I hear through social media, when I attend conferences or organise workshops, suggest to me change is happening in different countries and different practices.

For someone who cares about equity and everyday science learning these seem to be exciting times. Makerspaces such as the Mothership HackerMoms project set up by parents to reconfigure practices of participation centred on gender equity and childcare disrupt established patterns of interaction and access to technology (Dawson, 2017; O'Sullivan, 2018). Citizen science groups that meaningfully co-develop projects with refugee youth no longer seem like a distant dream (I can think of a couple in London run by community groups and the work of the French group L'Atelier des Jours à Venirs group takes this approach too) (L'Atelier de Jour à Venir, 2018). At the national scale, a collaboration between the Wellcome Trust and the UK government called Inspiring Science made a significant investment in UK science centres focused on diversifying their audiences. Activist networks of practitioners such as the Museum Detox group in the UK or the Museums as Sites for Social Action (MASS Action) project in the US are raising the profile of equity issues across the practices they touch. These are all profoundly hopeful practices.

My work has left me in the peculiar position of being both cynical and greedy. Cynical because I worry equity and inclusion have become a fashion in everyday science learning. And the problem with fashions is they rarely endure. As Feinstein (2017) argues, organisations such as museums face pressures that compete with and can undermine equity efforts, not least financial pressures. Thus, as we saw in the UK in the late 2000s, staff roles related to inclusion and community development were the first to be hit by the redundancies that resulted from austerity politics following the financial crash of 2007–2008. Unless equity and inclusion are at the heart of everyday science learning, they will remain at risk of seeming like luxury add-ons, all too swiftly dropped when money is tight or when a shiny new fashion comes along. When Gloria from the Afro-Caribbean group asked me how much difference taking part in this research project would make to inclusivity and equity in everyday science learning, I had to tell her I was not sure.

We must also take note of the right-wing politics growing in countries across the world. Our societies face serious challenges, not least that in some places racism, class discrimination, sexism and other forms of oppression are far from subtle. In 2017, staff at the Smithsonian National Museum of African American History and Culture in the US found a noose left in a gallery (Lopez, 2017). A noose. Take a moment to think about that. Another museum colleague, this time in the UK, recently told me racist and sexist abuse seemed to be growing in their visitor feedback and comments, on evaluation forms as well as on their website. At the very least then, organisations involved in everyday science learning must find a way to speak back to the structural inequalities that continue to shape and influence them, since no response looks alarmingly like institutional support for racism, class discrimination, sexism, other forms of oppression and their intersections.

My feelings of cynicism are closely related to my feelings of greed. Greed, because I want to write about the 'most' we could do rather than the 'least'. Greed, because I want so much more than to be able to list the few people, places, projects and organisations that centre their work on equity and inclusion organisations that, in doing so, appear somewhat radical against the backdrop of 'standard' everyday science learning practices. I want people involved in everyday science learning to learn from some of the practices of organisations such as the Black Cultural Archives in London and Glasgow Women's Library (Black Cultural Archives, 2018; Glasgow Women's Library, 2018). I want an equivalent of the Guerrilla Girls for everyday science learning as well as science, technology, engineering and maths (Guerilla Girls, 2018). I want equity-oriented change across the different fields involved in everyday science learning that does more than reflect a passing trend for politics and social justice, but that drives equitable change within and beyond these practices. I want these changes to be embedded, resilient, meaningful, widespread and sustainable.

Although I have explored issues of exclusion and equity in everyday science at length in this book, a series of uncomfortable questions remains in my mind, drawing on the work of people like Ortega (2006), Jennings and Jones-Rizzi (2017) and Puwar (2004). How can we build long-term, trusting relationships with the people and communities who hold the expertise to support community-centric practice? Can we relinquish enough control over content and practice that changes intended to support inclusion and equity are real, deep and sustainable? Can we acknowledge that we do not own scientific ideas, practices or the objects that embody these ideas? Can we, instead, own our ignorance and look for help?

And let's not forget our histories and the legacy of oppression we operate within – are we trustworthy? As Elizabeth Rasekoala (2018) put it, "do not trust the naked man who promises you clothes". Can we make the changes needed to make everyday science learning equitable without appropriating the knowledges, practices, time and effort of partners whose help is sorely needed? How can we alter practices radically enough that a them/us, insider/outsider dynamic is transformed?

These are difficult, thorny questions. Addressing them means being active, being brave, taking a stance, building relationships, being patient and being humble. Everyday science learning holds vast potential for disrupting rather than reproducing social disadvantages. But the kinds of changes I am writing about, within and across everyday science learning fields and, if we are to be ambitious, within science and technology writ large, as well as our societies, are no mean feat. I recognise the idealism inherent in writing about change in this way but I find comfort in the oft-quoted words of Lorde who wrote "revolution is not a one time event. It is being always vigilant for the smallest opportunity to make a genuine change in established, outgrown responses" (1984, pp. 140–141). From this perspective, each attempt to develop meaningfully inclusive, equitable practice, however small, helps us to learn more and embrace this challenge. What's more, from this perspective we can also see a way to think carefully about how everyday science learning practices could be used to support, leverage and campaign to work against the structural inequalities that shape our societies and the opportunities of those within them. I am still learning. And there is so much still to learn.

Note

1 Questions of generalisability, transparency and reliability in qualitative research are much discussed in terms of methods. Indeed, as discussed in Chapter One, these can be useful questions to ask, not least to understand clearly the terms under which research is carried out. My point here is somewhat different. It is that these kinds of questions can, in my experience, be used to draw attention away from the experiences of racialised groups, especially those experiencing multiple and intersecting structural inequalities, who are written off as extreme outliers.

References

Achiam, M., & Holmegaard, H. T. (2017). Informal science education and gender inclusion. In L. S. Heuling (Ed.), *Embracing the other: How the inclusive classroom brings fresh ideas to science and education* (pp. 32–40). Flensburg: Flensburg University Press.

American Alliance of Museums. (2018). *Facing change: Insights from the American Alliance of museums' diversity, equity, accessibility, and inclusion working group.* Arlington, VA: American Alliance of Museums.

American Anthropological Association. (2018). *Race: Are we so different?* Retrieved from www.understandingrace.org/about/overview.html

Ballard, H. L., Dixon, C. G. H., & Harris, E. M. (2017). Youth-focused citizen science: Examining the role of environmental science learning and agency for conservation. *Biological Conservation, 208,* 65–75.

Bang, M., Curley, L., Kessel, A., Marin, A., Suzukovich, E. S., & Strack, G. (2014). Muskrat theories, tobacco in the streets, and living Chicago as Indigenous land. *Environmental Education Research, 20*(1), 37–55. doi:10.1080/13504622.2013. 865113

Benhabib, S. (2002). *The claims of culture: Equality and diversity in the global era.* Princeton, NJ and Oxford: Princeton University Press.

Black Cultural Archives. (2018). *Home page.* Retrieved from https://blackculturalar chives.org

Dancstep, T., & Sindorf, L. (2018). Creating a female-responsive design framework for STEM exhibits. *Curator: The Museum Journal, 61*(3), 469–484.

Dawson, E. (2017). Social justice and out-of-school science learning: Exploring equity in science television, science clubs and maker spaces. *Science Education, 101*(4), 539–547. doi:10.1002/sce.21288

Dent, T. (2016). *Feeling devalued: The creative industries, motherhood, gender and class inequality.* (PhD Monograph). Bournemouth: Bournemouth University.

Dixon, C. A. (2012). Decolonising the museum: Cité Nationale de l'Histoire de l'Immigration. *Race & Class, 53*(4), 78–86. doi:10.1177/0306396811433115

Dixon, C. A. (2016). *The "othering" of Africa and its diasporas in Western museum practices.* (PhD Monograph), University of Sheffield, Sheffield.

Feinstein, N. W. (2017). Equity and the meaning of science learning: A defining challenge for science museums. *Science Education, 101*(4), 533–538. doi:10.1002/ sce.21287

Fraser, N. (1990). Rethinking the public sphere: A contribution to the critique of actually existing democracy. *Social Text,* (25/26), 56–80.

Fraser, N. (2003). Social justice in the age of identity politics: Redistribution, recognition, and participation. In N. Fraser & A. Honneth (Eds.), *Redistribution or recognition? A political-philosophical exchange* (pp. 7–109). London and New York: Verso.

Garibay, C. (2009). Latinos, leisure values, and decisions: Implications for informal science learning and engagement. *The Informal Learning Review, 94,* 10–13.

Garibay, C. (2017). *Metasynthesis of front-end studies excerpt, Winter 2017.* Chicago: Garibay Group.

Garibay, C., & Huerta Migus, L. (2014). *The inclusive museum: A framework for sustainable and authentic institutional change.* Chicago: The Association of Science-Technology Centres.

Garibay, C., Lannes, P., & González, J. (2018). *Latino audiences: Embracing complexity.* San Francisco: Exploratorium.

Gibson, L. (2001). *The uses of art: Constructing Australian identities.* St. Lucia: University of Queensland Press.

Glasgow Women's Library. (2018). *Home page.* Retrieved from https://womenslibrary. org.uk

Golding, V. (2009). *Learning at the museum frontiers: Identity, race and power.* Farnham and Burlington: Ashgate Pub. Co.

Gonsalves, A., Danielson, A., & Pettersson, H. (2016). Masculinities and experimental practices in physics: The view from three case studies. *Physical Review Physics Education Research, 12*(2), 1–15.

Guerilla Girls. (2018). *Home page.* Retrieved from www.guerrillagirls.com

Hahn, C. N. (2016). *The political house of art: The South African National Gallery 1930–2009.* (PhD Monograph), Goldsmiths College, University of London, London.

Jennings, G., & Jones-Rizzi, J. (2017). Museums, white privilege and diversity: A systematic perspective. *Dimensions,* 63–74.

Johnson, A., Brown, J., Carlone, H., & Cuevas, A. K. (2011). Authoring identity amidst the treacherous terrain of science: A multiracial feminist examination of the journeys of three women of color in science. *Journal of Research in Science Teaching, 48*(4), 339–366.

Kim, E.-J. A., Asghar, A., & Jordan, S. (2017). A critical review of traditional ecological knowledge (TEK) in science education. *Canadian Journal of Science, Mathematics and Technology Education, 17*(4), 258–270. doi:10.1080/14926156.2017.1380866

L'Atelier de Jour à Venir. (2018). *Activities: A diversity of approaches.* Retrieved from www.joursavenir.org/activities

Lawler, A. (1996). Goldin puts NASA on new trajectory. *Science, 272*(5263), 800–803.

Levin, A. K. (Ed.). (2010). *Gender, sexuality and museums.* London and New York: Routledge.

Longino, H. E. (1990). *Science as social knowledge: Values and objectivity in scientific inquiry.* Princeton, NJ: Princeton University Press.

Lopez, G. (2017). *Someone left a noose at the National Museum of African American History and Culture.* Retrieved from www.vox.com/identities/2017/6/1/15724126/noose-smithsonian-black-history-museum

Lorde, A. (1984). *Sister outsider.* Berkeley: Crossing Press.

Medin, D. L., & Bang, M. (2014). *Who's asking? Native science, western science, and science education.* Cambridge, MA and London: MIT Press.

Ong, M. (2005). Body projects of young women of color in physics: Intersections of gender, race, and science. *Social Problems, 52*(4), 593–617.

Ortega, M. (2006). Being lovingly, knowingly ignorant: White feminism and women of color. *Hypatia, 21*(3), 56–74. doi:10.1111/j.1527-2001.2006.tb01113.x

O'Sullivan, E. (2018). Excellence in the maker movement. *Journal of Peer Production, 3*(12), 46–50.

Philip, T. M., & Azevedo, F. S. (2017). Everyday science learning and equity: Mapping the contested terrain. *Science Education, 101*(4), 526–532. doi:10.1002/sce.21286

Puwar, N. (2001). The racialised somatic norm and the senior civil service. *Sociology, 35*(3), 651–670. doi:10.1017/S0038038501000335

Puwar, N. (2004). *Space invaders: Race, gender and bodies out of place.* Oxford and New York: Berg.

Rasekoala, E. (2018). *Social inclusion, equity and diversity: Time for change.* Paper presented at the Ecsite pre-conference, Geneva, Switzerland.

Saha, A. (2018). *Race and the cultural industries.* Cambridge: Polity Press.

Sandell, R. (2007). *Museums, prejudice and the reframing of difference.* London and New York: Routledge.

Sandell, R., Dodd, J., & Garland-Thomson, R. (2010). *Re-presenting disability: Activism and agency in the museum.* Abingdon and New York: Routledge.

Schiebinger, L. (2007). Getting more women into science: Knowledge issues. *Harvard Journal of Law and Gender, 30,* 365–378.

Science Museum of Minnesota. (2018). *Race: Are we so different.* Retrieved from www.smm.org/race

Taylor, M., & O'Brien, D. (2017). "Culture is a meritocracy": Why creative workers' attitudes may reinforce social inequality. *Sociological Research Online*, *22*(4), 27–47. doi:10.1177/1360780417726732

Yalowitz, S., Garibay, C., Renner, N., & Plaza, C. (2013). *Bilingual exhibit research initiative: Institutional and intergenerational experiences with bilingual exhibitions.* Washington, DC.: Centre for the Advancement of Informal Science Learning.

Yamada, M. (2015/1981). Asian Pacific American women and feminism. In C. Moraga & G. Anzaldúa (Eds.), *This bridge called my back* (pp. 68–72). Albany: State of New York University Press.

Young, I. M. (1990). *Justice and the politics of difference.* Princeton, NJ: Princeton University Press.

Young, I. M. (2000). *Inclusion and democracy.* Oxford and New York: Oxford University Press.

Appendix
Research methods

Social research is a glorious thing. Chapter One contains a two-paragraph summary of research methods I used to generate the data this book is based upon. I realise, of course, that many people will not wish to read anything more about my research methods. I confess however, that I am someone who will flick to the back of a book in the hopes of finding more about the details that make social research what it is. This is what you can find here, the hows and whys of which path was followed, my successful (and failed) attempts to generate and understand data. The tricky balance of carrying out research with people who ultimately become friends, but whose lives you only partially share.

Research design

Taking an ethnographic approach to this study was essential since it quickly became clear that parachuting into community settings to run a couple of focus groups would get me nowhere. Participant recruitment involved a catch-22 problem. Unsurprisingly, people who did not participate in everyday science learning were also not very interested in a research project about everyday science learning. Following a snowball approach, where one community gatekeeper introduced me to another and so on, I negotiated access to over 42 different grass-roots community groups.

My approach to participant recruitment was based on the descriptive data discussed in Chapter Three about who might not participate in everyday science learning and was designed to be exploratory rather than representative (Gobo, 2004). I began looking for neighbourhoods where people from both racialised groups and socio-economically disadvantaged backgrounds might live, in order to reach out to community groups who might be interested in taking part in a research project on everyday science learning, and I started with my own neighbourhood, which I described in Chapter One.

Of course group membership, whether organised around ethnic, socio-economic or other shared practices or identities, is not necessarily the identifying characteristic it is sometimes presumed to be (Hoggett, 1997; Spencer, 2006). As such, I did not assume the experiences of individual participants would or could represent the experiences of their community or people in similar social positions. Instead I just wanted to learn more about why people did what they did.

After an initial six months of contacting people, meeting community gate-keepers and visiting groups, I began to spend time with those who invited me back. Eventually, after several false starts, people from five community groups agreed to take part in the research and I began to meet them regularly. During the fieldwork period I spent days, evenings and weekends attending community events, visiting participants' houses and joining them at community centres to learn about them and how they saw science and everyday science learning.

As Rosemary McKechnie noted "ethnography entails taking note of much that does not seem immediately relevant, through getting to know people, and how they relate to each other and oneself" (1996, p. 129). Thus pages and pages of my field notes had nothing to do with science or everyday science learning. Indeed, these were topics more conspicuous by their absence than anything else. Over time my presence at community sites became more accepted, so alongside activities that were clearly research focused, I babysat participants' children and helped out at community festivals. I also began to ask participants more directly about science and everyday science learning, since these were not subjects that emerged without prompting. For busy people with fascinating lives, science and science-related activities were subjects of considerable silence. This was what led me to carry out the semi-structured focus groups and the majority of the interviews, as well as the four accompanied visits to science museums and a science centre.

The participants

Five community groups ultimately took part in the research I discuss in this book: an Afro-Caribbean group ($n = 7$), a Somali group ($n = 6$), a Sierra Leonean group ($n = 16$), a Latin American group ($n = 17$) and an Asian group ($n = 13$). Participants ranged from 18 years old to 76 years old and across the 59 participants, 41 were female and 18 were male.

These groups were grass-roots community groups who coalesced around a common sense of shared cultural heritage and 'race'/ethnicity. Participants were mixed in terms of age and educational backgrounds. Between one and five participants in every group were educated to degree level (including one participant with an environmental sciences MSc in the Sierra Leonean group, the highest science qualification across the five groups). In each group, however, some adults had no formal qualifications and had not been to school. Participants in every group had children, whom I met during fieldwork (but were not included in interviews or focus groups and though they were included in the accompanied visits their data were not recorded).

As I argued in Chapter One, categorising people is not a value-free mechanism; it creates and labels specific kinds of publics, whether as experts, expected publics or Others. Notably, participants' status as immigrants to the UK complicates neatly describing them in terms of 'race'/ethnicity and class. Participants experienced comparative marginalisation (in terms of access to labour markets, education, culture and politics) in the UK that resulted from their migration trajectories, colonialism and racist structural inequalities (Sassen, 2001; Vertovec, 2004). No participants were born in the UK, though around half had lived in the UK for a significant part of their lives (10 years or more for many and for some 30 to 40 years).

Although the combination of migration to the UK and structural inequalities created conditions for living in relative poverty speaks to the enduring legacies of colonialism and racism, this does not mean it is straightforward to describe participants as socio-economically disadvantaged or working-class (Bhopal, 2018; Gilroy, 2002). For instance, almost all participants sent money to family members living in diaspora or 'home' countries; many described their backgrounds as middle class at 'home'. In the UK however, due to a combination of devalued 'foreign' qualifications, limited English language fluency and the structure of the labour market, participants were unemployed or employed in precarious, badly paid jobs as temporary nurses, cleaners or security guards during the project.

It is important to understand 'race'/ethnicity and class as inextricably linked to the context of migration for participants and alongside the other intersecting subjectivities of their lives. In other words, while racism and class discrimination were the two most salient forms of structural inequality in participants' interactions with everyday science learning practices, along with gender, it is important to consider these issues as interwoven. I make this point here and from a more theoretical perspective in Chapter Two because I use 'race'/ethnicity and class throughout the analysis discussed in this book. I am, however, wary that these can be read and heard as analytically distinct themes in ways that risk whitewashing participants' experiences of class discrimination as though such experiences were not enmeshed in racism and colonialism, an issue I discuss further in Chapter Eight.

Analysis

Data were analysed, as is common in ethnographic work, at the same time as more data were generated (Hammersley & Atkinson, 1997). This meant I could follow up on these analyses within and across groups while I was still doing fieldwork. I followed a qualitative, constant comparative approach alongside divergent case analysis whereby themes (patterns, counter-examples, absences, contradictions and so on) were identified and refined, after which theories were mapped onto the data (Miles & Huberman, 1994). This approach meant each theme reflected the nature of the data, rather than simply the theoretical framework or only a small portion of the data (Hammersley & Atkinson, 1997; Silverman, 2001).

Chapters Three, Four, Five, Six and Seven contain illustrative extracts from field notes, interviews, focus groups, emails or audio-recorded accompanied visits. These extracts do not represent the most extreme examples of what I found out from this project, but rather are illustrative examples, used here to exemplify an aspect of participants' experiences. I make this point here because I want to be clear: when you read about Abdou's negative view towards museums or about how hard it was for Hamiido to use an interactive science exhibit, these were not one-off events, but something common across the five groups involved in the study.

I discussed the possibility of participatory analysis with people in each of the five groups. One participant from three of the groups (the Somali group, the Latin American group and the Sierra Leonean group) carried out data analysis alongside me, discussing their findings and my own and operating as a form of member-checking for the how I made sense of their stories (Charmaz, 2005). Although many participants were clear about not wanting to be involved in data analysis, I greatly valued these opportunities to think through data with participants. I am from a mixed European, white, middle-class background, with a science degree and a career in everyday science learning. This was evident enough to participants that the elderly Sierra Leonean women called me 'the white girl'. Participants in every group clearly saw me as an insider to science and everyday science learning. So it became important to me to talk through my interpretations of their words and lives with participants, even if we did not always end up agreeing (Duneier, 1999).

The most useful description of my position during fieldwork was that of "betweenness" (Nast, 1994, p. 57). Just as I was different in particular ways to the people who chose to work with me, we also had things in common, as is inevitable in any relationship. For instance, the politics of fieldwork for this project was such that with two groups we used parts of the data generated for their own ends. In one case I helped with evaluating a community festival and in another writing a funding application for a new mini-bus. Thus as a researcher I was in an in-between position, since the extent to which I was an insider or outsider was always conditional, relational and never absolute. This has advantages and disadvantages of course. On the one hand, I still get emailed photographs of people's grandchildren. On the other hand, drawing fieldwork to an end and moving on to new projects was difficult, as is perhaps inevitable with any ethnographic research.

References

Bhopal, K. (2018). *White privilege: The myth of a post-racial society.* Bristol: Polity Press.

Charmaz, K. (2005). Grounded theory in the 21st century: Applications for advancing social justice studies. In N. K. Denzin & Y. S. Lincoln (Eds.), *The Sage handbook of qualitative research* (pp. 507–536). London and Thousand Oaks, CA: Sage.

Duneier, M. (1999). *Sidewalk.* New York: Farrar, Strauss and Giroux.

Gilroy, P. (2002). *There ain't no Black in the Union Jack* (2nd ed.). Abingdon: Routledge.

Gobo, G. (2004). Sampling, representativeness and generalizability. In C. Seale, G. Gobo, J. F. Gubrium, & D. Silverman (Eds.), *Qualitative research practice* (pp. 435–456). London, Thousand Oaks, CA and New Delhi: Sage.

Hammersley, M., & Atkinson, P. (1997). *Ethnography* (2nd ed.). London and New York: Routledge.

Hoggett, P. (1997). *Contested communities: Experiences, struggles, policies.* Bristol: Policy Press.

McKechnie, R. (1996). Insiders and outsiders: Identifying experts on home ground. In A. Irwin & B. Wynne (Eds.), *Misunderstanding science? The public reconstruction of science and technology* (pp. 126–151). Cambridge: Cambridge University Press.

Miles, M. B., & Huberman, A. M. (1994). *Qualitative data analysis* (2nd ed.). London, Thousand Oaks, CA and New Delhi: Sage.

Nast, H. J. (1994). Women in the field: Critical feminist methodologies and theoretical perspectives. *The Professional Geographer, 46*(1), 54–66.

Sassen, S. (2001). *The global city: New York, London, Tokyo* (2nd ed.). Princeton, NJ and Oxford: Princeton University Press.

Silverman, D. (2001). *Interpreting qualitative data: Methods for analysing talk, text and interaction* (2nd ed.). London, Thousand Oaks, CA and New Delhi: Sage.

Spencer, S. (2006). *Race and ethnicity: Culture, identity and representation.* Abingdon and New York: Routledge.

Vertovec, S. (2004). Migrant transnationalism and modes of transformation. *International Migration Review, 38*(3), 970–1001.

Index